Studies in the History of Philosophy of Mind

Volume 19

AF801015

More information about this series at http://www.springer.com/series/6539

Marc Champagne

Consciousness and the Philosophy of Signs

How Peircean Semiotics Combines Phenomenal Qualia and Practical Effects

Springer

Marc Champagne
Department of Philosophy
Trent University
Peterborough, ON, Canada

Studies in the History of Philosophy of Mind
ISBN 978-3-030-10357-6 ISBN 978-3-319-73338-8 (eBook)
https://doi.org/10.1007/978-3-319-73338-8

Softcover re-print of the Hardcover 1st edition 2018

Printed on acid-free paper

This Springer imprint is published by Springer Nature
The registered company is Springer International Publishing AG
The registered company address is: Gewerbestrasse 11, 6330 Cham, Switzerland

To Serge, Henry, and Ahti, for believing in my work, in Alexander Bain's sense of belief.

Acknowledgments

I want to thank the following people for reading the manuscript and offering helpful feedback: Jake Beck, Tyler James Bennett, Vincent Colapietro, Chris Green, Henry Jackman, David Jopling, Kalevi Kull, Cathy Legg, Christine McKinnon, and David Skrbina (naturally, I bear sole responsibility for any mistakes or errors in the book). Kristin Andrews and David Stamos gave me good tips in the early stages of this project. Jelena Issajeva organized a stimulating symposium on the book at the 2015 Toward a Science of Consciousness Conference. Ahti-Veikko Pietarinen enabled a productive year of research at the University of Helsinki. My colleagues at Trent University fostered a supportive work environment. I want to thank David Chalmers for his words of encouragement. I also want to thank Henrik Lagerlund and Mikko Yrjönsuuri for curating this exciting series in philosophy of mind. Parts of this book have appeared, in different versions, in the following venues:

Champagne M (2016) Can pragmatists believe in qualia? The founder of pragmatism certainly did… Cybernetics and Human Knowing 23(2): 39–49

Champagne M (2015) Poinsot versus Peirce on merging with reality by sharing a quality. Versus 120: 31–43

Champagne M (2014) Referring to the qualitative dimension of consciousness: Iconicity instead of indexicality. Dialogue: Canadian Philosophical Review 53(1): 135–182

Champagne M (2013) Choosing between the long and short informational routes to psychological explanation. Philosophical Psychology 26(1): 129–138

Champagne M (2012) Russell and the Newman problem revisited. Analysis and Metaphysics 11: 65–74

Champagne M (2009) Explaining the qualitative dimension of consciousness: Prescission instead of reification. Dialogue: Canadian Philosophical Review 48(1): 145–183

Contents

Chapter 1
Introduction

Abstract In this introductory chapter, I preview how philosophy of signs can advance current debates about consciousness. Many philosophers have urged us to distinguish between what an experience does and what an experience feels like. This distinction seems sensible enough, but it renders scientific inquiry insufficient, insofar as experimental methods can only track an experience's detectable effects. However, Charles Sanders Peirce, the polymath who coined the term "qualia," saw no tension in this function/quality distinction. Peirce was the founder of American pragmatism and a professional scientist, so he clearly valued tangible verification. Still, his philosophical system countenanced intrinsic qualities under the heading of "Firstness." Since qualia are usually considered to be at odds with pragmatic verification, I set out to understand how Peirce could (or thought he could) have it both ways. The key, I suggest, is to see how we can sometimes insert distinctions between features that are always bound together. Such "prescissive" distinctions are essential to understanding how signs work. So, while semiotics is currently the least known branch of philosophy ending in –ics, a better understanding of that branch can remedy much of the current puzzlement about qualia.

> The elements of every concept enter into logical thought at the gate of perception and make their exit at the gate of purposive action; and whatever cannot show its passports at both those two gates is to be arrested as unauthorized by reason.
>
> Charles Sanders Peirce, "Pragmatism as the Logic of Abduction," lecture delivered at Harvard on May 14, 1903 (Reprinted in Peirce 1998, 241)

> Each *quale* is in itself what it is for itself, without reference to any other.
>
> Charles Sanders Peirce, "Quale-Consciousness," unpublished notes, circa 1898 (Reprinted in Peirce 1931–58, 6.224)

M. Champagne, *Consciousness and the Philosophy of Signs*, Studies in the History of Philosophy of Mind 19, https://doi.org/10.1007/978-3-319-73338-8_1

This book engages with a philosophical worry that the qualitative or phenomenal dimension of consciousness may not be captured by regular scientific explanation, that is, by the sort of inquiry which tracks tangible effects. There is a sense in which I agree and disagree that phenomenal consciousness poses such a problem, so I draw on the semiotic ideas of Charles Sanders Peirce to articulate in what respects I think the worry ought and ought not to be taken seriously. I think that, if we can come to see how this book's two starting quotations are consistent, we will have made progress in philosophy of mind.

Most aspects of human cognition are answerable to regular techniques of scientific investigation. Response times, skin moisture levels, eye movements, verbal reports, brain scans, blood flow rates—all can be tracked in controlled experiments that give a robust indication of what a subject thinks. This coverage seems quite sufficient. After all, "[i]f an experience is reported, or accessed in some weaker sense, then we can explain it by reference to how the brain thinks about itself. But if an experience happens without being accessible, then it becomes a phantom" (McDermott 2007, 518). However, not everyone agrees. In his 1974 paper "What Is It Like to Be a Bat?," Thomas Nagel argued that consciousness has an elusive "what it's like" component that is left out by regular scientific explanation. Science does a great job of figuring out how sense organs work, brains operate, and organisms respond. But, as Nagel points out, "in discovering sound to be, in reality, a wave phenomenon in air or other media, we leave behind one viewpoint to take up another, and the auditory, human, or animal viewpoint that we leave behind remains unreduced" (1974, 445). His suggestion, then, was that no amount of functional input-output correlations will add up to a quality.

Nagel suggested this during a "wave of reductionist euphoria" (1974, 435), so it took a while for his arguments to persuade others. Nagel's (2012) positive views never managed to attract many adherents, but as Paul Churchland writes, his early paper became a "prominent flag around which much antireductive opinion has rallied" (1996, 196). Roughly two decades after Nagel's piece appeared, Francisco Varela (1996, 331) would speak of an "outburst" of research on consciousness. A topic once considered taboo is now one of the most vibrant in philosophy.

Following Nagel, philosophers of mind like David Chalmers and Ned Block have argued (from different angles) that, even if one were to fully describe how a creature responds to stimuli, those input-output patterns would still leave out the quality of the relevant experience(s). Yet, before one can assess this claim of incompleteness, one has to get clear on what aspect is supposedly left-out. In order to make their grievance against functionalist explanation persuasive, these philosophers have introduced a distinction between "phenomenality" and "accessibility" (to use the jargon favoured by Block 1995). The concept of phenomenality is meant to capture how a conscious episode *feels*, whereas accessibility tracks what it *does*. Given this distinction, one needs a further step to say that an account of doing supplies one with an account of feeling. Many accounts of the mind tie consciousness to a property that is more manageable, like attention (see Prinz 2012), but I agree that "[t]o analyze consciousness in terms of some functional notion is either to change the subject or to define away the problem" (Chalmers 1996, 105).

This tension between experiential qualities and functional relations is not the traditional mind-body problem. The mind-body problem takes root in René Descartes. For that reason, it is, perhaps inevitably, tied to idealism and philosophical scepticism. It conceives the mind as an immaterial substance for which the only evidence is private introspection. To appreciate how this differs from the more recent philosophical problems that will concern me, consider Frank Jackson's "knowledge argument":

> Mary is a brilliant scientist who is, for whatever reason, forced to investigate the world from a black and white room *via* a black and white television monitor. She specialises in the neurophysiology of vision and acquires, let us suppose, all the physical information there is to obtain about what goes on when we see ripe tomatoes, or the sky, and use terms like "red," "blue," and so on. [...] What will happen when Mary is released from her black and white room or is given a colour television monitor? Will she *learn* anything or not? It seems just obvious that she will learn something about the world and our visual experience of it. But then it is inescapable that her previous knowledge was incomplete. (Jackson 1982, 130)

This argument does not rely on any of the standard Cartesian themes in order to be gripping. Mary the neuroscientist is no sceptic, so she never doubts that there is an external world or that she has/is a body. In fact, in Jackson's story, she knows all there is to know about both. The problem of experiential qualities arises from the fact that, in Mary's case, her body and the world have not (yet) met in the right way. Hence, until that close encounter takes place, we can wonder whether Mary fully appreciates what it is like to see the colour(s) that are absent from her room. When she exits that room, she can point, quite publicly, to the colour(s) at hand. The question is whether her language, scientific or vernacular, can do all the work needed to convey this quality. One does not become a substance dualist by suspecting that language somehow falls short. The task, then, is to clarify that "somehow."

Ideally, a full account of the conscious mind should cover both qualia and their practical effects. Far from being a pessimist, Nagel saw this "as a challenge to form new concepts and devise a new method—an objective phenomenology not dependent on empathy or the imagination" (Nagel 1974, 449). In this book, I want to take up that challenge. Whereas Nagel assumed that "such a solution, if it exists, lies in the distant intellectual future" (1974, 436), I want to argue that we should look to the past, not the future, to find the "objective phenomenology" that Nagel calls for. Specifically, I believe that the materials needed to assemble a robust account of consciousness already exist in Peircean semiotics. My book will thus be part history of philosophy, part current philosophy, and—hopefully—part future of philosophy as well.

Parties involved in the debates about consciousness know quite well that "the word *quale* and its plural *qualia* were introduced into philosophy as technical terms precisely in order to capture that aspect of an experience that escapes the scrutiny of any natural science" (Hattiangadi 2005, 342). Few seem aware, though, that this term "qualia" was introduced by the scientist and founder of pragmatism, Charles Sanders Peirce (1998, 272). Qualia were thus part of semiotics from the get-go. In fact, in some places, Peirce calls them qualisigns: "A *Qualisign* is a quality which is a sign. It cannot actually act as a sign until it is embodied; but the embodiment has nothing to do with its character as a sign" (Peirce 1998, 291). My book is devoted to unpacking all that this statement means and entails.

> The core of Peirce's arguments combines two theses: the mind is a kind of semiosis; sign processes are extended within the spatiotemporal dimension, so that something physical has to instantiate or realize them. According to the first one, the mind is the nature of the sign-action. The second thesis asserts that signs cannot act unless they are spatiotemporally realized. Thus, if a sign is to have any active mode of being, it must be materially embodied (or, at least, it results from a previous operation with material signs). (Queiroz and Atã 2014, 283–284)

Sober theses like these should quell any fears that discussing phenomenal consciousness will take us away from the concrete realities of empirical science and daily life. Nevertheless, Peirce's philosophy reminds us that it is not mandatory to always have those concrete realities in view. We can forget them, deliberately, by focusing solely on the intrinsic character of an experience. When we do this, we realize why, in a qualitative sign, its "embodiment has nothing to do with its character as a sign" (Peirce 1998, 291).

Peirce began studying Richard Whatley's *Elements of Logic* when he was twelve years old (see Brent 1998, 48) and honed his skills for the rest of his life, inventing a complete system of logical graphs when he was 57 years old (Roberts 1973, 11). As a logician relying on symbolic and diagrammatic notation to catalog and evaluate inferences, Peirce knew that signs can embody meaning. But, as a philosopher, he also wanted to know "what must be true of signs in order for them to embody meaning" (Liszka 1996, 10). It was a view of inquiry as a collective enterprise that "led Peirce to open the dusty folios of the medieval schoolmen" (Colapietro 1989, 2) in order to further his study of signs. Inspired by Duns Scotus' stance on the Trinity, Peirce realized that, since all signs involve a triadic relation among a sign-vehicle, an object, and an interpretation, we can adopt three different perspectives on any meaningful phenomena: we can consider (1) the phenomenon itself, (2) what it stands for, and (3) what it stands for to. The first perspective, which Peirce aptly called Firstness, is what lands contemporary philosophy of mind into trouble. Nevertheless, Peirce held that we can artificially isolate a qualitative character from whatever practical effect(s) it may have.

True, Peirce's pragmatic maxim tells us that if the object of a given concept does not "conceivably have practical bearings" (Peirce 1992, 132), then we have no reason to credit our concept with having an object. Interestingly though, Peirce never shaved-off qualia. On the contrary, as Paul Livingston explains in his historical survey of consciousness debates, "[f]or Peirce, qualia (often used as cognate to 'qualities') were already the most basic constituents of the totality of sensory experience, the ground of what he called Firstness or immediacy" (Livingston 2004, 6).

As I will emphasize throughout this book, Firstness is a relation-free state revealed only by artificial analysis. Now, some naturalists (e.g., Ross and Spurrett 2004) think that such analyses should carry no weight in debates about the mind. Peirce was conversant with—and contributed to—advances in psychology and neuroscience (see Pietarinen 2006, 71–76), but he was never that kind of naturalist.

His reason was simple. Although our eyes can see material things like a white flag or vial of dilating mercury, they cannot see the relations that make those things significant. Because "signs in their constitutive being as [triadic] relations are invisible to sense [...]" (Deely 2009, 236), Peirce rightly saw that a study of signs cannot make do without abstract philosophical methods.

> As to the important experimental research of semiotics, by means of questionnaires, tests, eye-tracking, brain-imaging and so on, all such procedures importantly add to our general understanding of how signs, meanings, and references are processed by human beings and their brains and minds, in some cases by different groups of human beings. But such results can never hope to reduce the generality of signs to any mere sum of such individual processing. (Stjernfelt 2013, 106)

I think that, to be realistic, a theory of consciousness must begin with an experiential baseline that is complex (Käufer and Chemero 2015, 80–83). I nevertheless think that humans are endowed with a unique ability to adulterate this baseline by isolating qualities. Opposing this, Cohen and Dennett (2011) have recently claimed that experiential qualities cannot be separated from the cognitive functions they are embroiled in. One immediate reaction upon hearing this is: of course they can. After all, when I spit out a food item because it is bitter, I can distinguish the bitterness from the spitting. Naturally, a lot turns on what sort of distinction we are talking about. One idea floating around since at least Hirst (1959) is that there may be respectable ways of granting that some things are the same yet different. Unless one wants to dismiss this suggestion altogether, one must at least explain why very smart thinkers constantly return, under different guises, to what David Papineau (2002, 6–7) calls the "intuition of mind-brain distinctness." Peircean semiotics gives us a way to do justice to this intuition.

Peirce did not address the "hard problem" of consciousness as it is currently understood. Even so, I believe his philosophy of signs can help us achieve a more satisfactory theoretical account of "what it's like" to enjoy conscious experience. The key, I suggest, is to see how we can sometimes insert distinctions between features that are always bound together.

Peirce argued that, to contemplate qualia in their Firstness, one must artificially suppose that all causal and inferential relations are absent (as we shall see, latent similarity relations resist this supposition of absence). Using a medieval terminology, Peirce called this deliberate myopia "prescission" or "prescissive abstraction" (like Peirce 1998, 270fn, I will use those terms interchangeably). Prescission is crucial in semiotic inquiry, because it allows us to evince an asymmetrical subsumption in the triadic sign. We can suppose a sign-vehicle without an object, or a sign-vehicle-and-object link without an interpretation, but we cannot go in the opposite direction (Peirce 1992, 4–6). Additionally, prescission allows us to discern three kinds of sign-vehicles (qualities, occurrences, and regularities) and three ways that these sign-vehicles can be related to objects (by similarity, causation, and convention). As we shall see, the nature of a sign-vehicle constrains the sort(s) of referential relations it can have. Reference by convention needs to be conveyed by a regularity, reference by causation needs to be conveyed by occurrences, and reference by similarity needs to be conveyed by a quality. Each time, though, the more complex relations presuppose the simpler ones, so we bottom out at ineffable experiential qualities.

Because prescissive distinctions always leave their complex starting point untouched, there is no reason why cognitive science should do things any differently. A better understanding of prescission can nevertheless soothe the worry that the "intrinsic quality of experience" (Harman 1990) has somehow been "left-out" (Levine 1997).

Peirce died in 1914, but for decades, few knew what to make of his semiotic theory. As Ahti-Veikko Pietarinen writes, "Harvard University to which Peirce's literary remains were eventually deposited, had in its possession a monster easier to lock up than harness" (2006, 46n22). At Harvard, Clarence Irving Lewis wrote a dissertation on "The Place of Intuition in Knowledge," under the supervision of Peirce's friend, Josiah Royce. Upon being hired at Harvard in 1920, Lewis was tasked with cataloging the Peirce papers. The two years spent doing this left their mark. "'How could the given be intelligible to the mind if it were independent of its interpretive activity?' This is a question which Lewis would not solve to his satisfaction until much later when he read Peirce" (Dayton 2016). In addition to being one of the first to engage with Peirce's work in logic (Lewis 1918, 79–106), Lewis was one of the first to use Peirce's new word "qualia" (Lewis 1946, 188).

Lewis taught W. V. O. Quine in the early 1930s. Quine clearly read Peirce—he reviewed the second volume of Peirce's *Collected Papers* soon after it appeared (Quine 1933)—but as a behaviourist he did little to advance our understanding of consciousness. Even so, Quine had fruitful discussions with his colleague, the linguist Roman Jakobson, who began teaching at Harvard in 1949. Jakobson started publicly mentioning the ideas of Peirce as early as 1952 (see Jakobson 1971, 555–556). In June 1974, he told participants at the first congress of the International Association for Semiotic Studies that engaging with the work of Peirce would bear dividends (see Jakobson 1979; reprinted in Jakobson 1988).

Umberto Eco was at that congress in Milan, and he took Jakobson's advice seriously (Proni 2015). Prompted by his study of Peirce, Eco was one of the first philosophers to work into his theory the idea of a similarity-based signification that is neither symbolic nor indexical. Moving away from a purely conventional interpretation of interpretation helped to cement Eco's (Eco 1990) growing impatience with the implausible relativism inadvertently supported by his earlier work (Eco 1976). Given Eco's established prominence, his change of mind on the controversial topic of images (Eco 2000, see Polidoro 2015) invigorated Peircean scholarship. In fact, European scholars were (and arguably remain) the go-to people for insight into this American thinker (see Rydenfelt 2014). By looking at the qualitative dimension of consciousness from a Peircean semiotic vantage, my book is part of this ongoing effort to rehabilitate and apply the mature ideas of Peirce.

As it turns out, Peirce (1998, 334–335) wasn't nitpicking when he insisted that his stance differed crucially from the pragmatisms of James, Dewey, and others. As work on the Peirce manuscripts proceeds, the full extent of this difference is dawning on scholars. A lot of the received wisdom about Peirce that one finds in mainstream histories and textbooks is just plain wrong. Specifically, Peirce was never a "verificationist." On the contrary, he took a sophisticated stance that countenanced *potential* signification. So, whereas some scholars still try to discuss Peircean pragmatism without discussing Peircean semiotics, such a compromise is untenable, since "[h]is pragmatism is a theorem of his theory of signs" (Fisch 1986, 435).

Peircean semiotics has informed cognitive science (see Steiner 2013), but many of the technical notions developed by Peirce remain poorly known in mainstream debates. For instance, in addition to the acquaintance and description identified by Bertrand Russell (1910–11), "[a] third kind of signification exists" that "is barely glimpsed in formal semantics today" (Legg 2013, 17). This third kind of signification is iconicity. If, as I suspect, icons are needed in order to successfully refer to qualia, then it is predictable that qualia should appear puzzling.

How do we explain the mind's qualitative dimension? If experiential qualities escape functional description, how could we ever refer to them? What sort of place, if any, should we assign qualities in the world? I believe Peircean philosophy of signs can help us answer some of these questions.

Peirce developed a phenomenology independently of Husserl (Spiegelberg 1956; Stjernfelt 2007, 141–159). Yet, since quite a bit of doctoring needs to be done in order to yield the insights that are of interest to a philosophic study of signs, semiotic inquiry is very different from phenomenological description. In many ways, it is closer to logic. The foundational assumption of semiotics is that all signs, wherever they are found, involve a triadic structure (Fisch 1983). In a rare pedagogic moment, Peirce (1998, 170–171, 425–427) likened the relation involved in the action of signs to "giving," insofar as the very idea commits one to countenancing not only (1) that which is given, but also (2) that to which it is given and (3) that which gives. This example was later made popular by Russell (1998, 59), who picked it up from Royce—that friend of Peirce who, "[f]or some reason" that Russell could not discern, "always liked triadic relations" (Russell 1998, 68). As we will see, it is unclear whether Russell really understood the Peircean emphasis on triadicity. In any event, like "giving," the passage at play in a sign cannot be reduced to pairs, so the point is that nothing below three places will do.

Identifying the sign with a generic triadic relation lets semiotic inquiry roam across many disciplinary domains. Thomas Sebeok (1979, 64) thus called ecumenicalism the "distinctive burden" of semiotics. Elimination is the cardinal sin of such a research program. Indeed, "if semiotics is the science of signs, as the etymology of the word suggests, then it does not exclude any sign. If, in the variety of the systems of signs, one discovers systems that differ from others by their specific properties, one can place them in a special class without removing them from the general science of signs" (Jakobson and Pomorska 1983; quoted in Sebeok 1991, 77). One is of course free to return to a local inquiry after having adopted such a wide vantage, but one will then do so with a renewed understanding of where the human mind and its products fit in the grander scheme of things.

I believe that, to make sense of qualia, what is needed is the removal of complexity. In a prominent collection of essays on consciousness, James H. Fetzer claims that "systems are conscious when they have the ability to use signs of specific kinds and not incapacitated from the exercise of that ability" (Fetzer 2003, 303). The definition of the sign Fetzer uses—which he attributes to Peirce—is "something that stands for something (else) in some respect or other for somebody" (Fetzer 2003, 303; see also Fetzer 1988). This view ensures that anything lower than triadic relations cannot be regarded as telling us anything useful about consciousness. Fetzer's proposal is telling, because it is the opposite of what I aim to accomplish in this book.

There is no denying that conceiving minds as things for which things can stand for other things is a good way to approach the "easy" (or, more appropriately, "easier") problems of consciousness; like the ability to discriminate, categorize, react to environmental stimuli, and so on (Chalmers 1995, 200). However, philosophical worries about consciousness arise from the fact that such a theory of sapience or thinking will not amount to a theory of sentience or feeling. By dipping below the level that makes cognitive processing possible, I want to show how this demand for a theory of sentience can fit into—or, more precisely, be subsumed in—an enriched account of sapience.

One can prescind a simple quality amid the stream of consciousness, but one has to make sure that whatever one reports about those impoverished scenarios does not help itself to the very resources supposed absent. Unfortunately, mainstream debates took on selected Peircean ideas without grasping their full semiotic motivation. Thus, taking their lead from Russellian philosophy of language, most philosophers of mind assume that causal reference is our only way to get in touch with the world and other minds. By dipping below the level of triadic relations, I want to approach phenomenal consciousness from a different—and more promising—angle.

Right now, semiotics (we owe the name to Locke) is arguably the least known branch of philosophy ending in –ics. I think that this is about to change. Nathan Houser, who spent decades editing Peirce's vast papers, surmises that, in the long run, "Peirce's semiotic may prove to be his most important contribution, really the creation of a new science" (in Bellucci et al. 2014, 129). Although I am unsure whether it should be considered a science, I am convinced that Peirce's triadic account of consciousness (Houser 1983) is exactly the sort of "fundamental theory" that David Chalmers has been searching for since his 1996 book *The Conscious Mind: In Search of a Fundamental Theory*. Let me now share the ideas that lead me to this conviction.

References

Bellucci F, Pietarinen A-V, Stjernfelt F (eds) (2014) Peirce: 5 questions. Automatic Press/VIP, Copenhagen

Block N (1995) On a confusion about a function of consciousness. Behav Brain Sci 18(2):227–287

Brent J (1998) Charles Sanders Peirce: a life. Indiana University Press, Bloomington

Chalmers DJ (1995) Facing up to the problem of consciousness. J Conscious Stud 2(3):200–219

Chalmers DJ (1996) The conscious mind: in search of a fundamental theory. Oxford University Press, Oxford

Churchland PM (1996) The engine of reason, the seat of the soul: a philosophical journey into the brain. MIT Press, Cambridge
Cohen MA, Dennett DC (2011) Consciousness cannot be separated from function. Trends Cogn Sci 15(8):358–364
Colapietro VM (1989) Peirce's approach to the self: a semiotic perspective on human subjectivity. State University of New York Press, Albany
Dayton E (2016) Clarence Irving Lewis. In: Internet encyclopedia of philosophy. Available via http://www.iep.utm.edu/lewisci/. Accessed 5 July 2016
Deely JN (2009) Basics of semiotics, 5th edn. Tartu University Press, Tartu
Eco U (1976) A theory of semiotics (trans: Osmond-smith D). Indiana University Press, Bloomington
Eco U (1990) The limits of interpretation. Indiana University Press, Bloomington
Eco U (2000) Kant and the platypus: essays on language and cognition (trans: McEwen A). Harcourt Brace and Company, New York
Fetzer JH (1988) Signs and minds: an introduction to the theory of semiotic systems. In: Fetzer JH (ed) Aspects of artificial intelligence. Kluwer Academic, Dordrecht, pp 133–161
Fetzer JH (2003) Consciousness and cognition: semiotic conceptions of bodies and minds. In: Smith Q, Jokic A (eds) Consciousness: new philosophical perspectives. Oxford University Press, Oxford, pp 298–322
Fisch MH (1983) Just how general is Peirce's general theory of signs? Am J Semiot 2(1–2):55–60
Fisch MH (1986) Peirce, semeiotic, and pragmatism. Indiana University Press, Bloomington
Harman G (1990) The intrinsic quality of experience. In: Tomberlin JE (ed) Philosophical perspectives, vol vol 4. Ridgeview, Atascadero, pp 31–52
Hattiangadi J (2005) The mind as an object of scientific study. In: Erneling CE, Johnson DM (eds) The mind as scientific object: between brain and culture. Oxford University Press, Oxford, pp 342–366
Hirst RJ (1959) The problems of perception. George Allen and Unwin, London
Houser N (1983) Peirce's general taxonomy of consciousness. Trans Charles S Peirce Soc 19(4):331–359
Jackson F (1982) Epiphenomenal qualia. Philos Q 32(127):127–136
Jakobson R (1971) Selected writings, volume II: word and language. Mouton, The Hague
Jakobson R (1979) Coup d'œil sur le dévelopement de la sémiotique. In: Chatman S, Eco U, Klinkenberg J-M (eds) A semiotic landscape. Mouton, The Hague, pp 3–18
Jakobson R (1988) A glance at the development of semiotics. In: Pomorska K, Rudy S (eds) Language in literature. Harvard University Press, Cambridge, pp 436–454
Jakobson R, Pomorska K (1983) Dialogues. MIT Press, Cambridge
Käufer S, Chemero A (2015) Phenomenology. Polity, Malden
Legg C (2013) What is a logical diagram? In: Moktefi A, Shin S-J (eds) Visual reasoning with diagrams. Birkhäuser, Heidelberg, pp 1–18
Levine J (1997) On leaving out what it's like. In: Block N, Flanagan O, Güzeldere G (eds) The nature of consciousness: philosophical debates. MIT Press, Cambridge, pp 543–555
Lewis CI (1918) A survey of symbolic logic. University of California Press, Berkeley
Lewis CI (1946) An analysis of knowledge and valuation. Open Court, La Salle
Liszka JJ (1996) A general introduction to the semeiotic of Charles Sanders Peirce. Indiana University Press, Bloomington
Livingston PM (2004) Philosophical history and the problem of consciousness. Cambridge University Press, Cambridge
McDermott D (2007) Dodging the explanatory gap—or bridging it. Behav Brain Sci 30(5–6):518
Nagel T (1974) What is it like to be a bat? Philos Rev 83(4):435–450
Nagel T (2012) Mind and cosmos: why the materialist neo-Darwinian conception of nature is almost certainly false. Oxford University Press, Oxford
Papineau D (2002) Thinking about consciousness. Oxford University Press, Oxford

Peirce CS (1931–58) The collected papers of Charles Sanders Peirce. Harvard University Press, Cambridge
Peirce CS (1992) The essential Peirce: selected philosophical writings, vol 1. Indiana University Press, Bloomington
Peirce CS (1998) The essential Peirce: selected philosophical writings, vol 2. Indiana University Press, Bloomington
Pietarinen A-V (2006) Signs of logic: Peircean themes on the philosophy of language, games, and communication. Springer, Dordrecht
Polidoro P (2015) Umberto Eco and the problem of iconism. Semiotica 206:129–160
Prinz JJ (2012) The conscious brain: how attention engenders experience. Oxford University Press, Oxford
Proni G (2015) Umberto Eco and Charles Peirce: a slow and respectful convergence. Semiotica 206:13–35
Queiroz J, Atã P (2014) Iconicity in Peircean situated cognitive semiotics. In: Thellefsen T, Sørensen B (eds) Charles Sanders Peirce in his own words: 100 years of semiotics, communication and cognition. De Gruyter, Berlin, pp 283–289
Quine WVO (1933) Reviews: collected papers of Charles Sanders Peirce. Isis 19(1):220–229
Roberts DD (1973) The existential graphs of Charles S. Peirce. Mouton, The Hague
Ross D, Spurrett D (2004) What to say to a skeptical metaphysician: a defense manual for cognitive and behavioral scientists. Behav Brain Sci 27(5):603–627
Russell B (1910–11) Knowledge by acquaintance and knowledge by description. Proceedings of the Aristotelian Society 11: 108–128
Russell B (1998) The philosophy of logical atomism. Open Court, Chicago
Rydenfelt H (2014) Peirce in Finland. European J Pragmatism Am Philos 6(1):51–57
Sebeok TA (1979) The sign and its masters. University of Texas Press, Austin
Sebeok TA (1991) Semiotics in the United States. Indiana University Press, Bloomington
Spiegelberg H (1956) Husserl's and Peirce's phenomenologies: coincidence or interaction. Philos Phenomenol Res 17(2):164–185
Steiner P (2013) C. S. Peirce and artificial intelligence: historical heritage and (new) theoretical stakes. In: Müller VC (ed) Philosophy and theory of artificial intelligence. Springer, Heidelberg, pp 265–276
Stjernfelt F (2007) Diagrammatology: an investigation on the borderlines of phenomenology, ontology, and semiotics. Springer, Dordrecht
Stjernfelt F (2013) The generality of signs: the actual relevance of anti-psychologism. Semiotica 194(1–4):77–109
Varela FJ (1996) Neurophenomenology: a methodological remedy for the hard problem. J Conscious Stud 3(4):330–349

Chapter 2
Calling on the Helpful Resources of Semiotic Inquiry

Abstract The name "semiotics" comes from John Locke, but the branch of philosophy that this name picks out remains mostly unknown in the mainstream literature on consciousness. This chapter will thus offer a primer on semiotics, both as an abstract inquiry and as an organized pursuit. The starting assumption of semiotic inquiry is that conventional meaning, inference from evidence, and resemblance-based representation all concern a single theoretical object, namely the sign. This is because the sign is a triadic relation that can be realized differently in different domains. Some of the bonds that sustain the action of signs are mind-dependent, whereas others are mind-independent. Although there are hints of this account in H. P. Grice's remarks about "meaning," the ecumenical perspective taken up by Peirce and other philosophers of signs can be traced back to Augustine. I recount this history and argue that Peirce was on the right track when he decided to approach the mind from a semiotic perspective.

2.1 Separating Qualities from Functions

Consciousness bundles together a lot of features, not all of which will concern me equally. So, like David Chalmers' book *The Conscious Mind*, I want to begin by zeroing in on a very narrow aspect:

> What is central to consciousness, at least in the most interesting sense, is *experience*. [...] According to the psychological concept, it matters little whether a mental state has a conscious quality or not. What matters is the role it plays in a cognitive economy. On the phenomenal concept, mind is characterized by the way it *feels*; on the psychological concept, mind is characterized by what it *does*. (Chalmers 1996, 3–4, 11)

According to Chalmers, cognitive science has a lot to say about the psychological side of the mind, but it has almost nothing to say about the phenomenal side. Chalmers (1996, 5) thinks it is presently hard to see what a fully developed theory of phenomenality would look like. Even so, he thinks that the phenomenal dimension of the mind is different enough from the psychological dimension to merit a distinct concept. I agree.

Unfortunately, the distinction in question has come to be associated with so-called "zombies." Zombies are often said to be "behaviourally indistinguishable"

M. Champagne, *Consciousness and the Philosophy of Signs*, Studies in the History of Philosophy of Mind 19, https://doi.org/10.1007/978-3-319-73338-8_2

(see Tanney 2004), but according to the construal of phenomenality endorsed by Chalmers (1996, 95), they would be psychologically indistinguishable as well. It is not just that one can imagine a creature whose sensory input and behavioural output functions map onto ours despite having a different physical realization; a sufficiently sophisticated robot passing the Turing test would qualify in that regard. Rather, the claim is stronger, since it pertains to the very information processing sandwiched between the inputs and outputs. It is widely understood that "[m]y zombie twin, for instance, has his eyes water just as I do when he eats too much Wasabi. Unlike me, […] [t]here is nothing it-is-like for him to taste Wasabi" (Majeed 2013, 252). One should keep in mind, though, that the broad notion of function also covers inferences. We may consider premises as input and conclusions as output. Hence, if a conscious person silently ponders xyz before doing P, her zombie counterpart would, under the same circumstances, also silently ponder xyz before doing P. So, as far as zombie-detection goes, Freud would be as impotent as Skinner.

Framed this way, the presence or absence of a qualitative dimension would be unverifiable. Philosophers like Paul Churchland have therefore argued that, if we want to charge a scientific account of the mind with coming up short, we should "endeavor to find in it some real *empirical* failing. Imaginary failings simply don't matter" (2005, 558). Although one might have expected a pragmatist to agree, the founder of pragmatism actually developed a set of subtle distinctions that lend some support to current contentions about qualia. Indeed, the function/quality distinction that Chalmers appeals to is appealing because it rests on a truth which Peirce's ordinal categories make clear: the idea of a relatum without a relation makes sense, but the idea of a relation without relata does not. This, in turn, permits an asymmetrical deletion. In developing a semiotic account of consciousness, I will make a host of adjustments, but I will never stray from this core insight.

If we read Chalmers' chapter on "Two Concepts of Mind" (1996, 3–31) closely, we notice that he first pinpoints a very narrow sense of "experience." Only once this is done does he ask us to conceive experience-free zombies. This means that, whatever stance one takes on the traditional issue of dualism versus (materialist) monism, the quality/function distinction can be rendered intelligible on its own grounds. Chalmers and his commentators seem to consider the distinction between the two concepts to be mere table setting, but I think it is where all the major action transpires. Chalmers insists that "[t]here should be no question of competition between these two notions of mind" (1996, 11). Call this the non-overlap thesis. He also insists that "[t]hey cover different phenomena, both of which are quite real" (ibid.). Call this the dual commitment thesis. The account of consciousness that I want to develop accepts the dual commitment thesis but rejects the non-overlap thesis.

Because "zombies" display no practical difference from regular humans, they violate the founding intent of Peirce's pragmatic maxim. Still, as we have just seen, before one can delete qualia from zombies, one has to focus on a very narrow construal that deletes all relations from a quality. As a result, prescission "becomes philosophically relevant when it is applied […] to disentangling the fundamental concepts which account for our experience and thought. In fact, it is for this purpose that Peirce clarified the method of prescision" (Gava 2011, 236). However, by comparison with pragmatism, Peirce's work on prescission is virtually unknown. It might therefore be helpful to distinguish two versions of Peirce.

There is, on the one hand, the better-known version who founded pragmatism and managed to get a few key articles published at an opportune time. Here, the functional side of the mind is played up. Yet, there is also, on the other hand, the lesser-known (but increasingly appreciated) version who made unparalleled advances in philosophy of signs which, until recently, remained mostly unpublished. Here, the non-functional side of the mind is given its due. This is a rough division, but it helps. The fact that the *Essential Peirce* collections are cut into two volumes renders that division even more vivid. The cover of the first volume (Peirce 1992) shows the younger Peirce, who instructed us to focus on the practical effects of a concept in order to clarify its meaning; whereas the second volume (Peirce 1998) shows the older, more heavily bearded, Peirce, who relentlessly investigated a neglected branch of philosophy called semiotics.

The second essay of the second *Essential Peirce* is entitled "What is a Sign?" Naturally, this will strike any newcomer as a great place to start. The essay begins by noting that "since all reasoning is an interpretation of signs of some kind," its titular question is "most necessary" but also "very difficult" (1998, 4). A mere four lines in, Peirce tells us that "[i]t is necessary to recognize three different states of mind" (ibid.). The first state of mind that Peirce invites us to consider is that of a red feeling, and nothing else. The reader who, saddled with an unexamined folk semiotic theory, came in expecting a discussion of traffic signals, will likely wonder at this point whether Peirce has veered off topic. He has not. Peirce's goal is to evince the conditions for the possibility of sign-action. He wants to make a point: with a quality like red and just red, there can be no flow of consciousness. If one finds such a flow, the impetus must have come from a source different from the quality itself.

Implicitly, Peirce is making an additional point, just as important, which is that we can follow through with his invitation to consider a quality like red in complete isolation from anything else. Logically, it can be done. Peirce makes sure to emphasize that "nobody really is in a state of feeling, pure and simple" as he just described (1998, 4). Yet, he observes that "whenever we are awake, something is present to the mind, and what is present, without reference to any compulsion or reason, is feeling" (ibid.).

Those who have read only the early Peirce should therefore take note: Peirce is not a "verificationist" who denies that we experience qualia. On the contrary, he begins by telling us that qualia are the most fundamental constituents of any mind. Phenomenal qualities thus take centre-stage, long before practical effects enter the scene and the drama of cognition proper begins. When Peirce asks "What is a sign?," he begins by distinguishing states of mind. Applying philosophy of signs to philosophy of mind is thus perfectly natural. As Thomas Short explains: "[S]ince the human mind, according to Peirce, is constituted by semeiotic processes of a special type, it should be possible to use the concept of semeiosis to analyze consciousness, and that precludes using the concept of consciousness to analyze semeiosis" (1986, 105).

Like cognitive science, philosophy of signs has "a very long past but a relatively short history" (Gardner 1985, 9). The word "semiotics" was introduced by a modern philosopher, its etymology is ancient Greek, its theoretical underpinnings are medieval, and the inquiry began flourishing in contemporary times. Alas, at present, most philosophers of mind are not well-versed in semiotics. Let me now try to rectify this.

2.2 Philosophy of Signs' Long Past

John Locke introduced the word "semiotics" in the penultimate paragraph of his *Essay Concerning Human Understanding*, where he surmised that semiotic inquiry might "afford us another sort of logic and critic, than what we have been hitherto acquainted with" (1825, 550). However, it was only in the twentieth century that this ambitious project began to be collectively realized. As Paul Bouissac writes in his introduction to Oxford's *Encyclopedia of Semiotics*:

> Semiotics represents one of the main attempts—perhaps the most enduring one—at conceiving a transdisciplinary framework through which interfaces can be constructed between distinct domains of inquiry. Other endeavors, such as the unified science movement of the 1930s or cybernetics and general systems theory in the 1950s and 1960s, met with only limited success. By contrast, semiotics remains a credible blueprint for bridging the gaps between disciplines and across cultures. (Bouissac 1998, ix)

In the estimate of many, no one has done more to evince that blueprint than Peirce. Hence, like many semioticians, I hold that, at present, "[t]here is no such thing as non-Peircean (or non-Sebeokian) semiotics, just as there are no non-Einsteinian physics or non-Darwinian biology" (Kilpinen 2008, 217).

Semiotics, however, is bigger than Peirce. Charles W. Morris was in all likelihood the first to explicitly teach a university course in semiotics in Chicago in the 1930s (Sebeok 1991a, 75, 123). The inquiry gained further institutional form at a 1964 conference in Bloomington, Indiana, when scholars from varied fields rallied under a common semiotic banner (Rey 1984, 92). The International Association for Semiotic Studies (IASS) held its first world congress in 1974, publishing its proceedings ever since. Although one has to select them wisely, reliable encyclopaedias (Cobley 2010; Sebeok 1994) and textbooks (Deely 1990; Savan 1987) in semiotics are now available, and the inquiry attracts a growing number of scholars who publish in established peer-reviewed journals like *Semiotica, Sign Systems Studies, Semiotic Inquiry*, and *The American Journal of Semiotics* (see the reliable sources catalogued in Champagne 2014b). Training-wise, there is a belief—expressed by the Russian cultural theorist Juri Lotman and subsequently defended by Thomas Sebeok (Deely 2009, 484)—that "semiotics is a field that one should not begin with" (Kull et al. 2009, xi). Thus, for better or worse, academic degrees in semiotics are awarded mainly at the graduate level (the Ph.D. program at UQÀM is the oldest, having been established in 1979).

Language has been a prominent theme in contemporary philosophy. If we follow Richard Rorty (1979) and Robert Brandom (1994) and construe "pragmatism" as an account of how communities do things with language, then absorbing the ideas of

Peirce in the analytic canon can seem like a natural move. Unfortunately, such an absorption would obscure more than it would reveal. Peirce never belittled language, but he was emphatic that human symbols are but one sort of sign among others. There may be legitimate reasons to limit a given philosophical inquiry to language. But, as Thomas Sebeok (1991b) brought out with sustained ferocity, one has to neglect or warp a lot of data in order to claim that language somehow marks a privileged boundary.

Somewhere along the way, many twentieth century thinkers took on a pair of alarmingly fallacious inferences (see Austin 2001, 30–31): words need not resemble their objects, therefore words never resemble their objects; and words (the species) never resemble their objects, therefore signs (the genus) never resemble their objects. These inferences have plagued both the analytic and continental traditions. Consider, for example, Ferdinand de Saussure's claim that conventional imputation is the glue that holds together the parts of a sign. If the Swiss linguist is right, then "[s]igns that are wholly arbitrary realize better than the others the ideal of the semiological processes […]" (De Saussure 2011, 68). This is a dogma. Although a study of linguistic symbols can definitely yield insights, such conventional signs are a proper subset of a much broader class of signs. Failure to recognize this leads to a fallacy of composition, wrongly assuming that a property of one or more of the parts of a whole is also true of that whole (Deely et al. 1986; Deledalle 2000, 100–113).

Because Saussure (2011, 16) saw the study of signs as a part of social psychology, his semiology had little or nothing to say about signs outside of culture. Louis Hjelmslev published a formal development of Saussure's project in 1943 (Hjelmslev 1969). By the time Roland Barthes met with the linguist A. J. Greimas in Egypt in 1949, Greimas informed him of (or created?) trends to come by telling him that "one cannot not know Saussure" (see Dosse 1997a, 68). Barthes later drew on the ideas of Saussure and Hjelmslev in his short "Éléments de sémiologie" (1977), which was widely used in literary circles. However, it was Maurice Merleau-Ponty who, in a prominent 1951 lecture (reprinted his 1964, 84–97), introduced Saussurean ideas into the philosophical mainstream. Ironically, Merleau-Ponty's pioneering work in embodied cognition ensured that arbitrariness and convention would never be the last word on meaning. French semiologists like Greimas have reluctantly had to acknowledge the presence of non-symbolic meaning (see the anecdote in Broden 2009, 577–578), but they have tended to relegate such events to phenomenologists (Ablali 2004). However, the add-ons came too late, as the relation-only view of French structuralism (Holdcroft 1991, 88–106, 119–130) eventually collapsed under the weight of its own internal contradictions (Dosse 1997b).

Compared with the social approach of semiology, semiotics has travelled down a very different path.

Instead of limiting signs to language and human customs, Thomas Sebeok (2001a, 10) held that "The criterial mark of all life is semiosis" and that "Semiosis presupposes life," so he allowed animals, plants, and even single cells to use signs (see El-Hani et al. 2006). Looking at the literature, "[t]he lowering of the semiotic threshold in semiotics during the last decades […] went parallel with the rediscovery of Peirce's broad concept of semiosis" (Nöth 2001, 15). As Sandra Rosenthal writes, Peircean semiotics "incorporates the dynamics of lived experience at its most rudimentary level, a dynamic that [...] grounds itself in those most rudimentary semiotic structures by which man experiences a world of appearing objects" (1979, 285). A biological turn may have recently begun in philosophy of mind (e.g., Thompson 2007), but it has been under way for decades in philosophy of signs (Sebeok 1988; Favareau 2010).

In a way, semiotic inquiry has returned to its roots, since the first signs to receive theoretical attention were medical symptoms (see the remarks by Marcel Danesi in Sebeok 2001b, xi–xvi). The following parallel can thus help to give a preliminary sense of semiotics' scope. In an attempt to iron out what he took to be an ambiguity in our common use of the word "mean," H. P. Grice called attention to the difference between sentences like "Those spots mean (meant) measles" and "Those three rings on the bell (of the bus) mean that the 'bus is full'" (Grice 1957, 377). Considering the first sentence, one cannot say "Those spots meant measles, but he hadn't got measles," whereas for the second sentence, it would make sense for one to say "But it isn't in fact full—the conductor has made a mistake" (Grice 1957, 377–378). One relation binds spots and measles, while another relation binds ringing bells and filled buses. Yet, the bell-bus relation is liable of being mistaken in a manner that the spot-measles relation is not, insofar as the effect that ringing bells have on people's conduct "must be something within the control of the audience, or at least the *sort* of thing which is within its control" (Grice 1957, 386). This led Grice to distinguish between what he called "natural meaning" and "nonnatural meaning."

Grice's distinction reflects lay usage. Indeed, Grice remarked, quite rightly, that "[t]his question about the distinction between natural and nonnatural meaning is, I think, what people are getting at when they display an interest in a distinction between 'natural' and 'conventional' signs" (1957, 379). Grice's work nevertheless left unanswered—or rather, unasked—a crucial question: why do we speak of "meaning" in both natural and nonnatural cases? One stock assumption is that this usage is a linguistic confusion which a more rigorous analysis could in principle redress. That is certainly one hypothesis. Yet, what if the kinship at hand is not a conflation but rather has a basis in fact? In his private lecture notes, Grice remarks that "Peirce uses the word 'sign' of a great variety of subjects" and wonders whether Peirce's use "draws attention to some important common feature which a great variety of items share" (in Pietarinen and Bellucci 2016, 86).

What we have in twentieth century philosophy of language are two sorts of significant relations, both placed under a single super-ordinate class of "meaning"—but the revolutionary implications of recognizing this kinship are not reaped (and even sometimes dismissed; as in Harman 1977). Philosophy of signs, by contrast, strives to explore that kinship and all it implies (Sebeok 2000).

As Umberto Eco explains in his widely-read book, *Semiotics and the Philosophy of Language* (1986b, 8), the starting assumption of semiotic inquiry is that conventional meaning (e.g., words), inference from evidence (e.g., symptoms), and resemblance-based representation (e.g., maps) all "concern a unique theoretical object," namely the "sign" writ large. This perspective rests on the realization—originating in the medieval period and later vindicated by Peirce—that all varieties of sign-action exhibit a common triadic structure which can and ought to be studied in its own right.

Semiotic theory thus augments the Gricean division in two respects. First, it adds something above the Gricean division, since it regards natural meaning and non-natural meaning as species of the genus "sign." Second, it adds something below the Gricean division, since it regards "natural meaning" as a genus with two species, namely indices and icons. Indices are "natural" in virtue of a mind-independent link of causality, whereas icons are "natural" in virtue of a mind-independent link of similarity.

Tracing the origins of this taxonomy takes us on a whirlwind tour of Western philosophy. One of the most interesting findings to have emerged from the work of Umberto Eco and his colleagues (Eco and Marmo 1989, 4–5) is the discovery, surprising at first, that Greek thought had no general notion of "sign" as we understand it today. In ancient Greek culture, we find on one hand the *semeion*, which, like a symptom, expresses a conditional association like "If the woman has milk, then the woman has given birth." This construal, which was part of the early development of medical science (Baer 1983; Baer 1988), also applied to subjects like meteorology, and was carefully discussed by the Stoics (see Eco 1986b, 29–33, 214–215; Manetti 1993, 97–110). The signification at play in the *semeion* rides on a correlation which would obtain with or without the inferential-like movement that finds in the manifest a trace of the hidden. This is the broad class of signs that natural scientists are usually interested in. The feature which allows interpretation to go from a sign-vehicle to an object is mind-independent. We can fail in our epistemic apprehension, but our misattribution leaves intact the factual link that would have otherwise secured it.

On the other hand, we find in Greek thought the *symbolon*, whose signification is wholly conventional, like insignia or flags. Etymologically, the word "symbol" comes from linked coins or clay-plates that were used to publicly announce marriages and contractual agreements (Eco 1986a, 153; Eco 1990, 9; Peirce 1998, 9). A steady correlation is involved in the *symbolon*, but there is nothing above and beyond the interpretation of a code that binds the relata. Because this imputed link is mind-dependent, it can "convey the nonexistent with a facility every bit equal to its power to convey thought about what is existent" (Deely 1990, 17). This allows human users to exploit channels not bound by constraints for truthfulness, going against a default biological hard-wiring in order to acquire a uniquely flexible cognitive resource (see Donald 2001). Symbols permeate culture, so they are the signs that social scientists are usually interested in.

Plato had famously argued in the *Cratylus* (1997, 101–156) that names originally bore a natural bond to their objects. But, more famously still, Aristotle critiqued this implausible view in *De Interpretatione*, insisting that a "*name* is a spoken sound significant by convention [...] because no name is a name naturally but only when it has become a symbol" (1984, 25).

In any event, what we have in Greek Antiquity is a division between nature and culture, reflected in the fact that there are different words for signs, not one (Manetti 2010b). This tacit taxonomy begins to change when we come to Augustine. For reasons that have nothing to do with philosophy proper, Augustine remained ignorant of Greek language throughout his productive life. So, when he felt the need to reflect on how the Christian God could speak to us through surface marks on a codex, he did not first verify what the Greeks had thought on the matter. Proceeding from his native Latin, he instead defined the *signum* generically as "something which is itself sensed and which indicates to the mind something beyond the sign itself" (Augustine 1975, 86; see also Perälä 2014, 362).

Whatever its shortcomings, this definition is novel, because it accommodates both natural and cultural correlations—the smoke that stands for fire and the white flag that stands for surrender (see Jackson 1969, 48–49). Thus, in the year 387, "Augustine unifies the two theories and the two classes of signs" (Manetti 2010a, 25). In doing so, Augustine laid the groundwork for a genuinely semiotic inquiry (Todorov 1992, 56–57; see Manetti 1993, 156–168). Other divides will linger, like the one between "formal" and "instrumental" signs (see Maritain 1959, 119–120; Furton 1995, 96–97). But, henceforth, two sorts of correlations—mind-independent and mind-dependent—will be recognized as falling under one super-ordinate class as signs *tout court*.

It may be surprising to find Augustine credited with inaugurating a model that would in time blossom into a sophisticated theory of representation. We may recall, for example, Wittgenstein's paragraphs at the outset of his *Philosophical Investigations* (2001, 2–3)—not exactly a work known for its historical scholarship—that depict Augustine as using names to merely "label" cognitively complete concepts. All the same, the consensus among scholars is that Augustine "affirms more strongly than earlier writers have done that words are merely one type of sign; this affirmation, which stands out with increasing sharpness in his later writings, is the cornerstone of the semiotic perspective" (Todorov 1992, 36; see Eco and Marmo 1989, 4–5; as well as Markus 1957). Recent years have thus witnessed a growing "body of research on the Middle Ages from the community of semioticians" that "now begins to amount to a subgenre of semiotic scholarship" (Evans 1987, 177). It is important to keep in mind, however, that "[Augustine] introduced to the Latins and to philosophy the sign as a theme, but he himself was never to thematize it" (Deely 2001, 218).

The covering model of the sign put forward by Augustine was disseminated in Peter Lombard's twelfth-century anthology of authoritative tenets, *Four Books of Sentences*. That widely-read anthology presented future generations of philosophers with a dilemma: is the unified kind "sign" a conflation or an insight? Views on the matter differed, but after Augustine it became mandatory to figure out what the common denominator between all these different signs might be.

A comprehensive answer to that question was developed by John Poinsot, a seventeenth century Spanish philosopher and theologian who left behind a difficult but ground-breaking *Tractatus de Signis* (2013). Poinsot was an heir to the long and intricate debate over signs that took Augustine's definition as its starting point. Some denied that there is truly a feature which unites the different kinds of signs (Deely 2004, 107). Poinsot, by contrast, tried to vindicate the original Augustinian proposal on principled grounds. Indeed, "it is clear that both Augustine (b. 354; d.

430) and John of St. Thomas [a.k.a. Poinsot] (b. 1589; d. 1644) were engaged in the same intellectual program and therefore belong together" (Gracia and Noone 2006, 1). Poinsot essentially engages with the cryptic but pregnant insight enunciated by his predecessor Thomas Cajetan: "A rose existing only in thought is not a rose, but a relation existing in thought is truly a relation" (quoted in Deely 1994, 22). In keeping with this, Poinsot argued that a sign, being a relation, is indifferent to whether it is found in the mind or in the world. Hence, its "being consists in relating, and this does not depend on us" (Rasmussen 1994, 410).

I will explore Poinsot's ideas more fully in Chap. 5. For now, it suffices to point out that Poinsot's proposal took place in one of the least-known periods in the history of Western philosophy, which consists of scholastic thought after René Descartes (Champagne 2008–09). Because Poinsot's theoretical contributions went almost unnoticed, we have to wait several centuries for late-medieval insights to be recovered and further articulated. The year 1690 nevertheless stands out, for this is when John Locke first gave "semiotics" its name at the end of his *Essay*. Locke, however, did not carry out the revolutionary promise of that project. As Peirce writes: "The celebrated *Essay Concerning Humane Understanding* contains many passages which [...] make the first steps in profound analyses which are not further developed" (1931–58, 2.649). Semiotics is one of them.

2.3 Peirce's Major Contribution

In the history of philosophy, big changes can result from little decisions. Bearing that in mind, it is worth noting that "the chapter proposing semiotic is commonly omitted from the many abridged editions of Locke's celebrated work that appear in succeeding centuries" (Deely 1981, 240). Although the work of Poinsot shows that "the doctrine of signs proclaimed by Locke did not have to wait 200 years to rise in the bosom of Peirce's complex and monumental work" (Santaella 1991, 155), philosophy of signs truly came of age with the work of Peirce. Showing an unusual historical sensitivity, Peirce connected anew with the medieval literature on signs that had been forgotten in the shuffle of modernity (see Beuchot and Deely 1995; Tiercelin 2006).

As a logician trained in framing things in the broadest terms possible, Peirce showed how any sign is perforce a three-place relation where something stands *for* something *to* something—regardless of what might fill these three place-holders on a given occasion. Delete any component of a semiotic triad and representation becomes impossible. A sign is therefore characterized, not by its material status, but by its relational structure:

> A Sign, or Representamen, is a First which stands in such a genuine triadic relation to a Second, called its Object, as to be capable of determining a Third, called its Interpretant, to assume the same triadic relation to its Object in which it stands itself to the same Object. The triadic relation is *genuine*, that is, its three members are bound together by it in a way that does not consist in any complexus of dyadic relations. (Peirce 1998, 272–273)

If one were to postulate that the basic categories of all things are Up, Down, Left and Right, this would not mean that the universe is populated with scattered ups, downs, lefts, and rights (that remain so irrespective of what is around them). Similarly, to be a First or a Third is to play a certain logical role. The roles themselves, moreover, can be switched around, so that when an interpretation is produced, it in turn can be interpreted. Peirce called this switching around "semiosis," the action of signs. He got this name from a book on inferences by the first century Epicurean philosopher Philodemus (1978; see De Lacy 1938; Manetti 2002). Peirce encountered this book while overseeing the work of a student at Johns Hopkins University (Marquand 1983).

Peirce took the Latin notion of sign to a new level of theoretical sophistication. In the course of his studies, Peirce came to hold in high regard the medieval writings of John Duns Scotus (see Boler 1963). Central to Scotus' position was a specific sort of separation which—as the scholastic catchphrase goes—is "more than nominal but less than real." Peter King summarizes it as follows: "The core intuition behind Scotus's formal distinction is, roughly, that *existential inseparability does not entail identity in definition*, backed up by the conviction that this is a fact about the way things are rather than how we conceive of them" (2003, 22; emphasis added). Scotus' formal distinction was part of a concerted defence (from early-Christian times onward) against accusations of polytheism:

> How can one reconcile the doctrine of the Trinity with a belief in the unity and simplicity of God? [...] The problems posed by the Trinity supplied the impulse for the development of the distinction [...]. Of course, it was not the *only* field of application, and the formal distinction came to be invoked in solving a host of purely philosophical problems. (Jordan 1984, 1)

Peirce wrote that, if Scotus' work was "adapted to modern culture, under continual wholesome reminders of nominalistic criticisms, I am convinced that it will go far toward supplying the philosophy which is best to harmonize with physical science" (1931–58, 1.6).

As it turns out, religious controversies gave birth to a technical tool well suited for the study of signs. One of Peirce's most important contributions to semiotic theory was his Scotus-inspired realization that, if one wants to rigorously and systematically unpack all that is implied by the misleadingly obvious notion of "sign," then one must recognize that every sign manifests both an unbreakable unity (as a sign whose significance is transparently given in a flash, as it were) and a multiplicity (as a step-by-step passage from a sign-vehicle to that which it re-presents). Construing any of the components that go into making representation possible as things somehow capable of existing without the collaboration of the others may not obliterate them, but it does rob them of the significance that allows them to serve useful cognitive functions. Thus, if we dissect a sign too crudely and start reifying the various parts we have uncovered, we ensure that those parts no longer have any representational value. *Mutatis mutandis*, construing the sign as some airtight atom which reflective thought cannot penetrate would drain all the properties that make it a sign.

Peirce first shared this result with the American Academy of Arts and Sciences, in an 1867 paper titled "On a New List of Categories." While not the most mature of Peirce's papers, that curt text presented a rich but forgotten way of approaching perennial questions of philosophy:

> The terms "prescision" [Peirce often changed the spelling] and "abstraction," which were formerly applied to every kind of separation, are now limited, not merely to mental separation, but to that which arises from *attention to* one element and *neglect of* the other. Exclusive attention consists in a definite conception or *supposition* of one part of an object, without any supposition of the other. Abstraction or prescision ought to be carefully distinguished from two other modes of mental separation, which may be termed *discrimination* and *dissociation*. Discrimination has to do merely with the essences of terms, and only draws a distinction in meaning. Dissociation is that separation which, in the absence of a constant association, is permitted by the law of association of images. It is the consciousness of one thing, without the necessary simultaneous consciousness of the other. Abstraction or prescision, therefore, supposes a greater separation than discrimination, but a less separation than dissociation. (Peirce 1992, 2–3)

Peirce probably learned about prescissive abstraction from a 1724 logic textbook by the English theologian Isaac Watts (see Peirce 1998, 541fn13). There may also be affinities with Kant (Gava 2011). In any event, Cary Spinks notes that "prescission is a difficult concept, but it is one of the most powerful developed by Peirce and also one of the few which he keeps throughout his life work" (1991, 23).

Prescission lets us distinguish features that are always bound (and that would hardly make sense without one another). A good way to explain this would be to liken semiotics to geometry. While one would be hard-pressed to find in the natural world a line with no girth or a point with no extension, we nevertheless have the ability to rigorously decompose any three-dimensional space and manipulate the dimensions it subsumes. The organization in such a case is not cardinal, but ordinal: nothing in a singular point entails a line or a volume, but the very notion of volume logically implies the line and the point. Semiotics is articulated around a similar insight. As Peirce showed, any representation involves a triadic relation that cannot be sundered; that is, one which cannot be reduced to the dyadic or the monadic on pain of no longer representing. To be meaningful, something (monadic quality) must stand for (dyadic relation) something else and be taken (triadic interpretation) as so standing. Nevertheless, using prescission, we can break these three dimensions down and recognize the specific semiotic role of each.

Peirce's failure to secure a place within the academic establishment effectively bequeathed to future generations the laborious task of understanding his unpublished writings. Thus, outside of a handful of influential papers on pragmatism written mainly in the 1870s (which he eventually repudiated), his later thought remained largely unknown.

> As regards Peirce's semiotic in particular, hardly anybody had paid any attention to it at all—it is clear from something [John] Dewey says in his correspondence with [Arthur F.] Bentley that, prior to the publication of [Charles W.] Morris's article on the foundations of the theory of signs, not even he had previously paid any real attention to that aspect of Peirce's thought [...]. At most, the term "semiotic" was thought of as referring to a crackpot scheme for classifying things called "signs" which nobody in philosophy had any interest in to begin with [...]. (Joseph Ransdell, quoted in Deledalle 2001, 220)

In principle, though, Peirce's mature ideas were available shortly after the First World War. His extensive letters to Victoria Lady Welby (Hardwick 1977), which dealt mainly with semiotics, were circulated in Europe and sent to prominent intellectual figures (like Bertrand Russell). Peirce's letters eventually reached C. K. Ogden, who was employed by Welby as a research assistant. Together with his co-author I. A. Richards, Odgen published excerpts of Peirce's letters in *The Meaning of Meaning* (1989, 279–290), a book which Charles Morris credits with identifying "the contours of a general theory of signs" (1971, 7).

The philosopher of mathematics Frank P. Ramsey, who translated Wittgenstein's *Tractatus* (2002) with Ogden, came to know of Peirce's ideas through this connection. In 1923, we find Ramsey arguing in a review of Wittgenstein's book that the Austrian genius would have benefited from "two words used by C. S. Peirce," namely "type" and "token" (Ramsey 1923, 468). Finally, after a lifetime of exclusion, some of Peirce's semiotic notions were being broadcast in a prominent forum, by a respected (if still emerging) Cambridge scholar, during the formative years of the analytic movement, in discussing what was to become one of its founding texts. Yet, regrettably, what could have been an occasion to connect with promising medieval conceptions simply became a newfangled jargon in which to reprise stale schemes. By the time anybody realized that the type/token distinction was supposed to be threefold, the error had been broadcast.

2.4 Tone-Deaf no More

David M. Armstrong, an advocate of materialism in philosophy of mind, wrote an introductory textbook that begins with the type/token distinction developed by "the great U.S. nineteenth-century philosopher, C. S. Peirce" (Armstrong 1989, 1). Armstrong produces a box with the word "the" inscribed twice in it and writes: "Peirce would have said that there were two tokens of the one type" (ibid., 2). This characterization, though not inaccurate, is incomplete, since the distinction developed by Peirce is in fact threefold. Armstrong got his "the the" example from this well-known passage from Peirce:

> There will ordinarily be about twenty *the*'s on a page, and of course they count as twenty words. In another sense of the word "word," however, there is but one word "the" in the English language; and it is impossible that this word should lie visibly on a page or be heard in any voice, for the reason that it is not a Single thing or Single event. It does not exist; it only determines things that do exist. Such a definitely significant Form, I propose to term a Type. A Single event which happens once and whose identity is limited to that one happening […] such as this or that word on a single line of a single page of a single copy of a book, I will venture to call a Token. (Peirce 1931–58, vol. 4.537)

However, the passage immediately continues with this sentence: "An indefinite significant character such as a tone of voice can neither be called a Type nor a Token. I propose to call such a Sign a Tone" (ibid.).

Many philosophers know this passage, but few cite it whole. Even Risto Hilpinen, in his Presidential Address to the Charles Sanders Peirce Society, truncates his quote (see Hiplinen 2012, 260). It is hardly surprising, then, that Linda Wetzel alludes to this passage in her book on types and tokens (Wetzel 2009, xi) yet completely neglects to mention the tone. The silent transformation of Peirce's trichotomy into a dichotomy is now virtually ubiquitous (see Hutton 1990; Guttenplan 1995, 596–597).

Peirce's (1998, 364) pioneering work in topology convinced him that triads have a special status (see the proof in Burch 1991), so it is particularly unfortunate to see his ideas denatured. I have already recounted how Ramsey (1923) failed to mention the tone in his review of Wittgenstein's *Tractatus*. More knowledgeable philosophers could have intervened to correct matters. John Dewey had studied briefly under Peirce at Johns Hopkins University in the 1880s, but it took a while before Dewey (1946) divulged in print what he had learned. So, apart from a few timid interventions (e.g., Williams 1936, 702), the propagation of Peirce's ideas was far from optimal. In a 1949 article, Arthur Burks—the editor of Peirce's *Collected Papers*—reassured the philosophical public at large that the full type/token/tone distinction was articulated by Peirce "in a way which is too bound up with his system of categories to be of use outside his philosophy [...]" (Burks 1949, 673). I want to change all that. John Boler notes that quality is "certainly the least clear of the categories, and the one that receives the least attention" (1963, 123). I think the latter part of this statement accounts for the former.

Although he misappropriates the type/token/tone distinction, Armstrong correctly remarks that it is "a perfectly general distinction applicable to any subject whatever" (1989, 1; for more on Armstrong and Peirce, see Legg 2001). Like many, however, Armstrong uses "type" and "token" as synonyms for "universal" and "particular." That is not correct. "How far are the basic categories of Peirce's phenomenology either particulars or universals? In describing Firstnesses as qualities of feeling Peirce never makes their status plain in terms of this disjunction. All he requests is the disregard of the question of reality and of connections with other phenomena" (Spiegelberg 1956, 174). The type/token/tone distinction thus belongs to semiotics, not metaphysics (not the metaphysics of Plato, at any rate). The distinction is instead "based on the idea that a given entity, assumed to be a sign, can be regarded in respect to any or all of three types of properties it has—monadic, dyadic, triadic (i.e., one-term, two-term, three-term)—depending upon the analytical needs in some concrete semiotic inquiry" (Sebeok 1994, 1130).

Signs, as we saw, are individuated by their triadic structure. If what we have in view is the whole triad, then we are at the level of what Peirce called Thirdness. In such a case, we have a relation between two relata grasped as a relation by some third thing beyond it. This is usually the level of interest, especially when one is studying some particular cultural or natural instance of sign-use. But, if our goal is a philosophical analysis of the representational structure itself, we may want to go further. If we suppose this relation between two relata as it would be without any further recognition of it as a relation, we are dealing with Secondness. Two and only two things are now involved, so we have effectively left the realm of intelligibility and entered that of brute contiguity (see Champagne 2014a). Prescinding still further, we may also want to suppose one of the relata without it entering into any relation with another. Peirce writes: "[T]he idea of a quality is the idea of a phenomenon or partial phenomenon considered as a monad, without reference to its parts or components and without reference to anything else" (1931–58, 1.424). If we do this, we eliminate whatever alterity allowed that relatum (the term now becomes a misnomer) to have a "contour." Thus, when we prescind relation away so as to consider only that which is related, we may no longer think of the resultant tone as we do a token, since doing so would require us to delimit it in some fashion and ascend back to Secondness. The prescissive analysis leaves its initial object of study untouched. But, if we choose to prescind all the way, Firstness is as far as we can go, and we obtain a quality that *could* be actualized but isn't.

Although the declarative intelligibility expected of introspective reports is possible solely by recourse to triadic representation, this appeal does not preclude but in fact presupposes simpler dyadic and monadic ones which are patently not beholden to any form of mediation (see Peirce 1992, 257–262). The more elementary states of Firstness and Secondness require the addition of a third term in order to be interpreted, but we can discern their ordinal priority.

To better appreciate these prescissive distinctions, we can look at the following example (taken from Sebeok 1994, 1130; adapted from Peirce 1931–58, 2.230):

> Because of his long *fast*, he was too weak to stand *fast* or hold *fast* or even to run *fast*.

The truncated type/token dichotomy would tell us that there are four tokens and one type of "fast," but the correct Peircean verdict is that "fast" appears three times as type, four times as token, and once as tone. Hence, the Peircean perspective I champion doesn't just name things differently, it counts things differently. If one were to prefer saying "twelve eggs" instead of "a dozen eggs," this change would be merely terminological. If, however, one were to prefer saying "a score of eggs" instead of "a dozen eggs," this change would carve the world differently. Adopting the type/token/tone distinction thus results in a difference that makes a difference, at least as far as our taxonomies are concerned.

The idea of token is perhaps the easiest to compass. To be a token is to be an occurrence, something that has a discrete spatial and temporal location. In contrast, neither the type nor the tone is bound by such immanence. The tone is a quality—considered prior to its occurrence as token. If we prescind, we can isolate the qualitative feature that is common to the tokens "fast" and "fast." To be sure, this quality—in this case a configuration of marks—is very much there as constitutive of each token, and there is no way for us to get to "suchness" except through a "thisness." But, the tone itself enjoys a priority which enables us to logically isolate it while disregarding its numerically distinct manifestations. As an ordinal First, the tone "fast" is merely a potentiality, a "something" that could be employed to stand for something else (but doesn't have to be). If and when such a quality occurs, "fast" is a Second. It can now be used to indicate a specific spot on a page. To the extent that such an occurrence is not a singleton but appears repeatedly in a way that is not merely haphazard (e.g., a habit), "fast" is a Third.

It is important to stress that "[a] class, of course, is neither a tone nor a type in Peirce's sense" (Willard 1983, 284), because a class can have random members whereas a type generates tokens according to a repeatable rationale. This recurrence is what allows a type to sustain a conventional association. Children learning a natural language may be adept at inducing meanings from a poverty of stimulus (Chomsky 1980), but they still need a minimum of two examples plus an open-ended possibility of implementing a learned rule again. There is no such thing, for example, as a word that can be used only once. Of course, as a token, a word can (only) be used once. However, what makes a brute event like a noise capable of bearing conventional meaning is that it can be re-cognized and re-employed. Even so, the first time one hears a spoken word, that sound has a distinctive tone (as would pure gibberish). There is thus a tremendous difference between a type that does occur repeatedly and a tone that could occur repeatedly. In order not to miss this crucial nuance, Peirce saw fit to use different names.

Although the notion of "tone seems very close to what many subsequent writers have meant by 'type'" (Willard 1983, 284), the type/token/tone distinction tracks an important nuance that the coarser-grained type/token distinction misses. Consider the following situation. A set of things—orange traffic cones, say—are placed before a knowing subject. The subject, noticing that there is some feature(s) in common, then proceeds to tease out that common feature(s). A story can thus be told about how the shared colour or shape of the traffic cones was "abstracted away" from those immanent particulars. Yet, whatever its merit, solubility, or outcome, this set-up has already skipped over what interests me. Indeed, nothing prevents one's investigation from starting with a single item instead of a group of items. After all, a single traffic cone will have a quality (colour, shape, etc.) even if no other cones like it exist. It may therefore be helpful to replace our earlier sentential example with the following:

fast

Looking at this single token, the two theses I am concerned with are that 1) the here-and-now "thisness" of the singular occurrence can be prescinded from its qualitative "suchness," and 2) the "suchness" or tone one arrives at by prescission determines what similar tokens would be like without indicating whether such other tokens in fact exist. Going from a plurality of tokens to unencountered tokens may be a (risky) induction, but going from a single token to unencountered tokens is an abduction—a (riskier) surmise that other tokens with this tone might exist. Umberto Eco is philosophically impressed by the fact that, in perceptual judgement, "*that* yellow is like *all* the yellows I have seen" (2000, 115). But, to my mind, we should be more impressed by the fact that, even on the first occasion (Mary exiting the cave, say), that yellow will be like all the yellows one will ever see. One will skip right over this nuance if one begins with a collection of tokens.

In keeping with this, Peirce warned that, when we talk of abstracting out a common feature, we often clump together two logically distinct moments. To abstract a common feature, one must first pull out that feature from each individual and then treat what one has pulled out as if it could exist in this detached form. The first step is prescission and the second step is a reification Peirce called "hypostatic abstraction." Because these are different mental operations, Peirce urges us to "relieve the stem 'abstract' from staggering under the double burden of conveying the idea of prescission as well as the unrelated and very important idea of the creation of an *ens rationis* [...]" (1998, 352). Hypostatic abstraction "furnishes us the means of turning predicates from being signs that we think or think *through*, into being subjects thought of" (Peirce 1931–58, 4.549). Peirce is right that this can be a crucial step. However, I argue that, in philosophy of mind, it is crucial that one not perform this additional step, otherwise one will be left with a thing-like quale that empirical evidence will seek but never find.

A colour like orange may be prescinded, but orange does not exist apart from orange things. Peircean semiotics thus sits token occurrences in the middle of all of its systematic classifications, because "if a sign is to have any active mode of being, it must be materially embodied" (Queiroz and Atã 2014, 284). As David Savan explains, a tone "can not be encountered as such in experience [...]. What this means is that the empirical student of semiotics must use Peirce's trichotomy (if he uses it at all) as an analytical tool, by means of which to distinguish three different aspects of semiosis [...]. Empirically, no sign belongs exclusively to one of these classes" (1987, 23–24). Like a Russian doll, a type always presupposes tokens, and a token always presupposes a tone. Even if it makes little sense to think of "fast" as existing in only one of these respects, be it a quality that never occurs or a regularity that never manifests itself, prescission allows us to carefully peel off these three layers.

Like semiotic inquiry itself, the type/token/tone distinction employed by Peirce has a long history. For instance, "what distinguishes Avicenna's treatment of *essence* is the way he distinguishes three ways of taking it: as existing in individual things and so determining their kind, as understood to be shared by many such things, and as it is in itself" (Gracia and Noone 2006, 199). Avicenna's analysis influenced Scotus. According to Scotus,

> The first act of the intellect is the immediate and simple apprehension of an individual in so far as it is present and existing, and, as such, the first act of the intellect is opposed to the second or abstractive act of the intellect which reaches the object in its essence. [...] This second act of the intellect gives us knowledge of the essence of an object considered in abstraction from existence, whereas the former act gives us knowledge of an object as existent and actually present. (Almeder 1973, 4–5)

Talk of priority can engender confusions, so I want to reiterate that, in ordinary lived experience, the token comes first. It is only afterward that prescissive thought can decide to disregard what gives a token its contrast/outline. Hence, the simple qualities countenanced by Peircean philosophy of signs "are the artificial product of a highly sophisticated analysis, and not genuine existents revealed to ordinary, everyday scrutiny" (Goudge 1935, 536). So, it becomes a fallacy of sorts to take a doctored product of our thinking to be a discovered fact independent of that intervention. This, on my diagnosis, is the root cause of the "hard problem" of consciousness.

Peirce was a student of history. But, in its aspiration to reinvent all wheels, twentieth-century philosophy butchered his type/token/tone distinction, because it could not recognize that distinction's pre-modern rationale. Indeed, the vocabulary of the parties to the present controversy over qualia and Peirce's triadic vocabulary do not easily match up. A crucial question for the former debate is: should we quantify over qualia? Nagel, Block, and Chalmers think that we should; whereas Dennett, Churchland, and others think that we should not. Yet, if most of our philosophical theories have until now been unable to adequately fathom the qualitative dimension of conscious life without running into all sorts of implausible consequences, perhaps this is because those theories have been trying to capture that object of study with dichotomies ill-suited to the task. This, at any rate, is the hypothesis I would now like to explore.

References

Ablali D (2004) Sémiotique et phénoménologie. Semiotica 151(1–4):219–240

Almeder R (1973) Peirce's pragmatism and Scotistic realism. Trans Charles S Peirce Soc 9(1):3–23

Aristotle (1984) The complete works of Aristotle, vol 1. Princeton University Press, Princeton

Armstrong DM (1989) Universals: an opinionated introduction. Westview Press, Boulder

Augustine (1975) De dialectica (trans: Jackson BD). Reidel, Dordrecht

Austin JL (2001) Truth. In: Lynch MP (ed) The nature of truth: classic and contemporary perspectives. MIT Press, Cambridge, pp 25–40

Baer E (1983) A semiotic history of symptomatology. In: Eschbach A, Trabant J (eds) History of semiotics. John Benjamins, Amsterdam, pp 25–40

Baer E (1988) Medical semiotics. University Press of America, Lanham

Barthes R (1977) Elements of semiology (trans: Lavers A, Smith C). Hill and Wang, New York

Beuchot M, Deely JN (1995) Common sources for the semiotic of Charles Peirce and John Poinsot. Rev Metaphys 48(3):539–566

Boler JF (1963) Charles Peirce and scholastic realism: a study of Peirce's relation to John Duns Scotus. University of Washington Press, Seattle

Bouissac P (ed) (1998) Encyclopedia of semiotics. Oxford University Press, Oxford

Brandom RB (1994) Making it explicit: reasoning, representing, and discursive commitment. Harvard University Press, Cambridge

Broden T (2009) The phenomenological turn in recent Paris semiotics. In: Deely JN, Sbrocchi LG (eds) Semiotics 2008. Legas Press, Toronto, pp 573–583

Burch RW (1991) A Peircean reduction thesis: the foundations of topological logic. Texas Tech University Press, Lubbock

Burks AW (1949) Icon, index, and symbol. Philos Phenomenol Res 9(4):673–689

Chalmers DJ (1996) The conscious mind: in search of a fundamental theory. Oxford University Press, Oxford

Champagne M (2008–09) What anchors semiosis: how Descartes changed the subject. Semio Inq 28–29(3–1):183–197

Champagne M (2014a) Just do it: Schopenhauer and Peirce on the immediacy of agency. Symposium 18(2):209–232

Champagne M (2014b) Semiotics. In: Pritchard D (ed) Oxford bibliographies in philosophy. https://doi.org/10.1093/OBO/9780195396577-0179

Chomsky N (1980) Rules and representations. Columbia University Press, New York

Churchland PM (2005) Chimerical colors: some phenomenological predictions from cognitive neuroscience. Philos Psychol 18(5):527–560
Cobley P (ed) (2010) The Routledge companion to semiotics. Routledge, London
De Lacy EA (1938) Meaning and methodology in Hellenistic philosophy. Philos Rev 47(4):390–409
De Saussure F (2011) Course in general linguistics (trans: Baskin W). Columbia University Press, New York
Deely JN (1981) The relation of logic to semiotics. Semiotica 35(3–4):193–265
Deely JN (1990) Basics of semiotics. Indiana University Press, Bloomington
Deely JN (1994) How does semiosis effect renvoi? Am J Semiot 11(1–2):11–61
Deely JN (2001) Four ages of understanding. University of Toronto Press, Toronto
Deely JN (2004) Iberian fingerprints on the doctrine of signs. Am J Semiot 20(1–4):93–156
Deely JN (2009) Semiotics and academe: at the heart of the problem of knowledge. In: Deely JN, Sbrocchi LG (eds) Semiotics 2008. Legas Press, Toronto, pp 476–493
Deely JN, Williams B, Kruse FE (eds) (1986) Frontiers in semiotics. Indiana University Press, Bloomington
Deledalle G (2000) Charles S. Peirce's philosophy of signs: essays in comparative semiotics. Indiana University Press, Bloomington
Deledalle G (2001) À la source de la sémiotique triadique. Sem Inq 21(1–3):211–227
Dewey J (1946) Peirce's theory of linguistic signs, thought, and meaning. J Philos 43(4):85–95
Donald M (2001) A mind so rare: the evolution of consciousness. W. W. Norton and Company, New York
Dosse F (1997a) History of structuralism, volume 1: the rising sign, 1945–1966 (trans: Glassman D). University of Minnesota Press, Minneapolis
Dosse F (1997b) History of structuralism, volume 2: the sign sets, 1967–present (trans: Glassman D). University of Minnesota Press, Minneapolis
Eco U (1986a) On symbols. In: Deely JN, Williams B, Kruse FE (eds) Frontiers in semiotics. Indiana University Press, Bloomington, pp 153–180
Eco U (1986b) Semiotics and the philosophy of language. Indiana University Press, Bloomington
Eco U (1990) The limits of interpretation. Indiana University Press, Bloomington
Eco U (2000) Kant and the platypus: essays on language and cognition (trans: McEwen A). Harcourt Brace and Company, New York
Eco U, Marmo C (eds) (1989) On the medieval theory of signs (trans: Kelly S). John Benjamins, Amsterdam
El-Hani CN, Queiroz J, Emmeche C (2006) A semiotic analysis of the genetic information system. Semiotica 160(1–4):1–68
Evans J (1987) Per speculum in aenigmate: signs of the word in modern semiotics. Poet Today 8(1):173–180
Favareau DF (2010) An evolutionary history of biosemiotics. In: Favareau DF (ed) Essential readings in biosemiotics: anthology and commentary. Springer, Dordrecht, pp 1–77
Furton EJ (1995) A medieval semiotic: reference and representation in John of St. Thomas' theory of signs. Peter Lang, Bern
Gardner HE (1985) The mind's new science: a history of the cognitive revolution. Basic, New York
Gava G (2011) Peirce's "prescision" as a transcendental method. Int J Philos Stud 19(2):231–253
Goudge TA (1935) The view of Charles Peirce on the given of experience. J Philos 32(20):533–544
Gracia JJE, Noone TB (eds) (2006) A companion to philosophy in the middle ages. Blackwell, Oxford
Grice HP (1957) Meaning. Philos Rev 66(3):377–388
Guttenplan S (ed) (1995) A companion to the philosophy of mind. Blackwell, Oxford
Hardwick CS (ed) (1977) Semiotic and significs: the correspondence between Charles S. Peirce and Victoria Lady Welby. Indiana University Press, Bloomington
Harman G (1977) Semiotics and the cinema. Q Rev Film Stud 2(1):15–24
Hiplinen R (2012) Types and tokens: on the identity and meaning of names and other words. Trans Charles S Peirce Soc 48(3):259–284

Hjelmslev L (1969) Prolegomena to a theory of language (trans: Whitfield FJ). University of Wisconsin Press, Madison

Holdcroft D (1991) Saussure: signs, system, and arbitrariness. Cambridge University Press, Cambridge

Hutton C (1990) Abstraction and instance: the type-token relation in linguistic theory. Pergamon Press, Oxford

Jackson BD (1969) The theory of signs in St. Augustine's De doctrina christiana. Revue d'études Augustiniennes et Patristiques 15(1–2):9–49

Jordan MJ (1984) Duns Scotus on the formal distinction. Dissertation. Rutgers University, New Brunswick

Kilpinen E (2008) Memes versus signs: on the use of meaning concepts about nature and culture. Semiotica 171(1–4):215–237

King P (2003) Scotus on metaphysics. In: Williams T (ed) The Cambridge companion to Duns Scotus. Cambridge University Press, Cambridge, pp 15–68

Kull K, Salupere S, Torop P (2009) Semiotics has no beginning. In: Deely JN (ed) Basics of semiotics. Tartu University Press, Tartu, pp xi–xxvii

Legg C (2001) Predication and the problem of universals. Philos Pap 30(2):117–143

Locke J (1825) An essay concerning human understanding. Thomas Davison, London

Majeed R (2013) Pleading ignorance in response to experiential primitivism. Philos Stud 163(1):251–269

Manetti G (1993) Theories of the sign in classical antiquity. Indiana University Press, Bloomington

Manetti G (2002) Philodemus' "De signis": An important ancient semiotic debate. Semiotica 138(1–4):279–297

Manetti G (2010a) Ancient semiotics. In: Cobley P (ed) The Routledge companion to semiotics. Routledge, London, pp 13–28

Manetti G (2010b) The inferential and equational models from ancient times to the postmodern. Semiotica 178(1–4):255–274

Maritain J (1959) Distinguish to unite, or the degrees of knowledge. Scribner's, New York

Markus RA (1957) St. Augustine on signs. Phronesis 2(1):60–83

Marquand A (1983) The logic of the Epicureans. In: Peirce CS (ed) Studies in logic by the members of the Johns Hopkins University. John Benjamins, Amsterdam, pp 1–11

Merleau-Ponty M (1964) Signs (trans: McCleary RC). Northwestern University Press, Evanston

Morris CW (1971) Writings on the general theory of signs. Mouton de Gruyter, The Hague

Nöth W (2001) Protosemiotics and physicosemiosis. Sign Syst Stud 29(1):13–27

Ogden CK, Richards IA (1989) The meaning of meaning. Harcourt, New York

Peirce CS (1931–58) The collected papers of Charles Sanders Peirce. Harvard University Press, Cambridge

Peirce CS (1992) The essential Peirce: selected philosophical writings, vol 1. Indiana University Press, Bloomington

Peirce CS (1998) The essential Peirce: selected philosophical writings, vol 2. Indiana University Press, Bloomington

Perälä M (2014) Psychology of language: ancient and early medieval theories. In: Knuuttila S, Sihvola J (eds) Sourcebook for the history of the philosophy of mind: philosophical psychology from Plato to Kant, Studies in the history of philosophy of mind, vol 12. Springer, Dordrecht, pp 359–378

Philodemus (1978) On methods of inference (trans: De Lacy PH, De Lacy EA). Bibliopolis, Naples

Pietarinen A-V, Bellucci F (2016) H. Paul Grice's lecture notes on Charles S. Peirce's theory of signs. Int Rev Pragmat 8(1):82–129

Plato (1997) Complete works (trans: Reeve CDC). Hackett, Indianapolis

Poinsot J (2013) Tractatus de signis: the semiotic of John Poinsot (trans: Deely JN, Powell RA). St. Augustine's Press, South Bend

Queiroz J, Atã P (2014) Iconicity in Peircean situated cognitive semiotics. In: Thellefsen T, Sørensen B (eds) Charles Sanders Peirce in his own words: 100 years of semiotics, communication and cognition. De Gruyter, Berlin, pp 283–289

Ramsey FP (1923) Critical notice of Tractatus logico-philosophicus. Mind 32(128):465–478
Rasmussen DB (1994) The significance for cognitive realism of the thought of John Poinsot. Am Cathol Philos Q 68(3):409–424
Rey A (1984) What does semiotics come from? Semiotica 52(1–2):79–93
Rorty R (1979) Philosophy and the mirror of nature. Princeton University Press, Princeton
Rosenthal SB (1979) C. S. Peirce: pragmatism, semiotic structure, and lived perceptual experience. J Hist Philos 17(3):285–290
Santaella L (1991) Review of John Poinsot, Tractatus de signis. J Specul Philos 5(2):151–159
Savan D (1987) An introduction to C. S. Peirce's full system of semeiotic. Toronto Semiotic Circle, Toronto
Sebeok TA (1988) In what sense is language a "primary modeling system"? In: Broms H, Kaufmann R (eds) Semiotics of culture. Arator, Helsinki, pp 67–80
Sebeok TA (1991a) Semiotics in the United States. Indiana University Press, Bloomington
Sebeok TA (1991b) A sign is just a sign. Indiana University Press, Bloomington
Sebeok TA (ed) (1994) Encyclopedic dictionary of semiotics. Walter de Gruyter, Berlin
Sebeok TA (2000) Semiotics as bridge between humanities and sciences. In: Perron P, Sbrocchi LG, Colilli P, Danesi M (eds) Semiotics as a bridge between the humanities and the sciences. Legas Press, Toronto, pp 76–100
Sebeok TA (2001a) Global semiotics. Indiana University Press, Bloomington
Sebeok TA (2001b) Signs: an introduction to semiotics. University of Toronto Press, Toronto
Short TL (1986) Life among the legisigns. In: Deely JN, Williams B, Kruse FE (eds) Frontiers in semiotics. Indiana University Press, Bloomington, pp 105–119
Spiegelberg H (1956) Husserl's and Peirce's phenomenologies: coincidence or interaction. Philos Phenomenol Res 17(2):164–185
Spinks CW (1991) Peirce and triadomania: a walk in the semiotic wilderness. Mouton de Gruyter, Berlin
Tanney J (2004) On the conceptual, psychological, and moral status of zombies, swamp-beings, and other "behaviourally indistinguishable" creatures. Philos Phenomenol Res 69(1):173–186
Thompson E (2007) Mind in life: biology, phenomenology, and the sciences of mind. Harvard University Press, Cambridge
Tiercelin C (2006) The importance of the medievals in the constitution of Peirce's semeiotic and thought-sign theory. In: Fabbrichesi R, Marietti S (eds) Semiotics and philosophy in Charles Sanders Peirce. Cambridge Scholars, Newcastle upon Tyne, pp 158–184
Todorov T (1992) Theories of the symbol (trans: Porter C). Cornell University Press, Ithaca
Wetzel L (2009) Types and tokens: on abstract objects. MIT Press, Cambridge
Willard D (1983) Why semantic ascent fails. Metaphilosophy 14(3–4):276–290
Williams DC (1936) Tokens, types, words, and terms. J Philos 33(26):701–707
Wittgenstein L (2001) Philosophical investigations (trans: Anscombe GEM). Blackwell, Oxford
Wittgenstein L (2002) Tractatus logico-philosophicus (trans: Ogden CK, Ramsey FP). Routledge, London

Chapter 3
Using Prescission and the Type/Token/Tone Distinction

Abstract In the previous chapter, I argued that Peirce was on the right track when he approached the mind from a semiotic perspective. Having offered a primer on semiotics, I now want to use some of those helpful resources. Ned Block distinguishes access-consciousness and phenomenal-consciousness. Convinced that his distinction is a real one, Block posits a module in the brain responsible for phenomenal experiences. However, I argue that we can make better sense of Block's distinction if we regard it as a prescissive one. In order to clarify this stance, I examine experiments conducted by George Sperling. Sperling designed tasks that let test subjects access visual experiences that they previously could not act upon. Block thinks that Sperling's results vindicate his claims, but strictly speaking Sperling's experiments tracked only access-consciousness. I thus conclude that, to establish that phenomenal experiences were present prior to being accessed, we must prescissively suppose the subjects' tasks absent.

3.1 Block's Real Separation of Phenomenon and Access

Those who think conscious life has a qualitative dimension usually gloss the situation in the following terms: there are token brain states on one side, there are token qualitative states on another side, and this results in a gap (Farrell 1950; Levine 1983). Early theorists like U. T. Place (1956) assumed that the sides in want of a union should be grouped into types. When Donald Davidson (1970) explicitly focused on tokens instead of types, philosophers were given a new option. Jerry Fodor (1974, 100) was one of the first to use the type/token distinction to describe this new position in philosophy of mind. In his first critique of the functionalist program, Ned Block (1978, 261) followed Fodor and took the type/token distinction for granted.

The type/token distinction has since become the unquestioned starting point for work in philosophy of mind (see for example Gozzano and Hill 2015, 1). Given the historical omission I outlined in Chap. 2, the idea of tone—construed as something irreducible to tokens or types—simply never comes up. I want to use this missing notion to open up an overlooked stance.

M. Champagne, *Consciousness and the Philosophy of Signs*, Studies in the History of Philosophy of Mind 19, https://doi.org/10.1007/978-3-319-73338-8_3

The philosopher of mind working with a dichotomous palette of types and tokens is forced to conclude that if phenomenal qualities are real enough to be discerned, they must exist as something genuinely distinct. Block (1995a), for example, urges philosophers of mind to countenance a qualitative "phenomenal-consciousness" which exists apart from a functional "access-consciousness." Of course, once this much has been granted, the need to relate the disparate sides (with "bridge laws," "neural correlates," and the like) surfaces with particular inevitability. Peirce's complete type/token/tone distinction allows one to bypass this. If it is true that "we find in Peirce no traditional philosophical arrangement that creates a mind-body problem" (Pietarinen 2006, 76), then we can use Peircean ideas to recuperate in a more plausible way the intuitions that Block appeals to.

Here is how Block motivates his distinction. The psychology literature often describes cases where a person is missing one or more of the aspect(s) and/or faculty(ies) we typically expect consciousness to have. The stock assumption is that understanding these abnormal cases can better our understanding of consciousness in its normal state. John Searle (1992, 107–108), for instance, argues that if epileptics in the grip of a seizure do not display any flexibility and creativity in their behaviour, then we can conclude that flexibility and creativity are important traits of consciousness. Block contends that "[a]lthough some variants of this sort of reasoning have some merit, they are often given more weight than they deserve, because of a persistent fallacy involving a conflation of two very different concepts of consciousness" (Block 1995a, 228). In order to put an end to this conflation (or at least ensure that it no longer goes unnoticed), Block introduces a distinction.

Block calls the functional side of the mind "access-consciousness" and the qualitative side "phenomenal-consciousness." Supposedly, there can be access without phenomenon (Block 1995a, 243–245) and phenomenon without access (ibid., 239–243). Block thus inserts a "real" distinction (in Scotus' sense). If Block's mutual-independence holds, then a missing functional response is not sufficient grounds to infer that a phenomenal aspect is also missing. Likewise, enjoying a phenomenal experience is not sufficient grounds to infer that the experience in question is being cognitively accessed. I do not endorse Block's mutual-independence, but since it gets many things right, it can act as a useful foil.

Let us consider the first situation, where agents presumably process information without actually enjoying any phenomenality. Block believes that there is support for this in the case of "blindsighted" persons. People who are blindsighted report being blind, yet they have reliable functional responses to stimuli when pressed to venture an answer (see Bornstein and Pittman 1992; Milner and Rugg 1992; Nelkin 1996). Referring to the information-processing model developed by Daniel Schacter (1989), Block (1995a, 229) suggests that these generally correct judgements (about unseen things) are made possible because the patients somehow bypass a hypothetical "phenomenal-consciousness module" and proceed straight to the executive system whose end-product is overt behaviour (e.g., decisions, actions, and utterances). That, as it stands, is a provocative suggestion. As Chalmers explains, "Block's idea suggests an implausible epiphenomenalism *within* the information-processing story. Indeed, if the module has no effect on other processes, then we could lesion it with no external change (same reports, even), and no empirical evidence could support the hypothesis" (1997, 149). As if this wasn't inflammatory enough, Block enjoins us to fathom a "super-blindsighted" person who would declare that "Now I know that there is a horizontal line in my blind field even though I don't actually see it" and for whom visual information "simply pops into his thoughts in the way that solutions to problems we've been worrying about pop into our thoughts" (1995a, 233).

Something is wrong here. Blindsighted persons are supposed to be a) able to see, in the sense of being capable of having the proper responses to visual stimuli put before them; and b) unable to see, in the sense that there is nothing "it is like" for them to perform (a). For that conjunction as a whole to obtain, one has to determine whether both conjuncts obtain. Yet, when we go back to the original clinical observations, we find that the only way (b) is ascertained is by asking the patients if they experience anything during the relevant acts, to which they answer in the negative. Unlike everywhere else in psychology, the introspective reports of blindsight patients get a free pass.

For example, a patient (called G. Y.) reported that his visual experiences were "like black on black." Black on black is not nothing—it even suggests a dim outline. Professional phenomenologists have rightly warned that untrained introspection produces skewed results (Zahavi 2007, 27–32). Despite this, researchers preferred to credit the patient with a complete lack of phenomenal content. Their only reason was that the patient "still insists that the use of visual terms is for lack of a better alternative because in fact he does not see the stimulus" (Stoerig and Cowey 1997, 554–555). All that underwrites the "in fact" is the patient's insistence. In another case, a patient (called D. B.) was asked whether a stick was horizontal or vertical. Following a forced-choice guessing paradigm, this patient was informed that he scored above chance and was asked: "Did you know how well you had done?" to which he replied "I didn't—because I couldn't see a darn thing" (exchange quoted in Weiskrantz 1986, 24). This informal conversation was deemed sufficient to establish that the subject's experiential lights were off. One wonders to what extent the follow-up query alleviated the methodological challenge at hand: "'So you *really* did not know you were getting them right?' 'No,' he replied, still with something of an air of incredulity" (Weiskrantz 1986, 24; emphasis added).

Discussing the alleged blindness of a patient who can spatially track any X shown to it, Block writes: "Temporarily taking his word for it, I am assuming that he has no P-consciousness of the X" (1995a, 233). It is unclear, though, whether Block's temporary assumption is ever replaced by anything more tangible. After all, success at tracking tasks suggests sight, not blindness. So, the only thing leading researchers to posit blindness is an introspective report, nothing more. Well, if simply taking a person's claims at face value is all there is to establishing the absence or presence of qualia, then I too can insist with great vigour that all is dark inside. If a researcher is prepared to record my statements, I could parallel park a car and we could publish those revolutionary findings. So, either blindsight is legitimate and the hard problem is not, or the hard problem is legitimate and blindsight is not.

Now, it has been suggested by Block (1995a, 233) that "blindsight" can also occur in monkeys. However, because a monkey cannot verbalize what it experiences, it is even more questionable to ascribe "blindsight" to such an animal. If an animal bumps into things, its failure to perform regular motor functions in response to visual inputs can be a solid indicator of blindness. The problem, though, is that the monkeys in question don't bump into things (see Stoerig and Cowey 1997, 549). In fact, "blindsighted" creatures (human or monkey) are supposed to function quite well. Therefore, to describe them as blind is to abandon the simplest explanation.

It is not as if other explanations are unavailable. In primates, stimulation of the eyes triggers optic nerves that then activate a portion in the back of the brain called the striate cortex. Researchers thus reason that a monkey lacking a striate cortex is a blind monkey. This seems like a plausible inference. Yet, in reality, total eradication of a given brain region is very difficult to achieve. "For example, there might be diffusion of light within the eye so that some of it spreads into the intact field and provides a kind of subtle cue about the presence of a visual event" (Weiskrantz 1986, 11). Philosophers of mind eager to get mileage out of such cases might want to gloss over the fact that, in the most famous case (the monkey Helen), "*almost* the entire visual cortex of the monkey's brain" (Humphrey 1992, 88; emphasis added) was surgically removed, but the question of residual tissue matters. By analogy, philosophers of biology would hardly declare kidneys unnecessary solely on account that, once, a patient had almost all of her kidney tissue removed yet still managed to filter some blood.

Even if we were to assume for the sake of argument that complete removal of a monkey's striate cortex has been histologically verified, our sense of amazement would stem from a contrast. If we believe that the striate cortex is needed for vision, we will understandably be surprised when the relevant input-output functions survive its removal. Monkey "blindsight" thus trades on sustaining a joint endorsement of two contradictory claims, namely that "Cortical activation is needed for vision" and "Cortical activation is not needed for vision." Which is it? If a brain region is considered necessary for the performance of a function, and that function is shown to occur even in the absence of the brain region, then the brain region can no longer be considered necessary for the performance of the function in question. The situation is analogous to removing a cornerstone crucial to holding a building erect and then finding out that the building does not collapse. We could either say that 1) the building can mysteriously stand in spite of the cornerstone's absence, or we could say that 2) the stone was not a cornerstone after all. Clearly, option (2), which says that it was not a cornerstone, is the best explanation.

Lawrence Weiskrantz, the neuroscientist who coined the term "blind + sight," remarks that his oxymoron quickly caught on (1986, v). In philosophy of mind, enthusiasm for these cases has been fueled in part by their resemblance to "zombies," those hypothetical creatures that perform all humans can without enjoying any of the qualitative experiences (Kirk 1974; Kirk 2005; Chalmers 1996, 94–105). To his credit, Block (1995a, 233) prudently states that he "doesn't know whether there are any actual cases of A-consciousness without P-consciousness." Owen Flanagan, however, believes that "the case of blindsight shows its actuality" (1992, 149). As we have just seen, there are good reasons to doubt such a claim.

Block's mutual-independence of access and phenomenon entails another possibility that is more plausible—as in the case of a busy person who "hears" but does not "notice" the loud drilling noise that has been present near her during an engaging conversation (Block 1995a, 234). In terms of Schacter's model, this would mean that the hypothetical phenomenal consciousness module would be activated yet have no repercussion on an executive system that could trigger reactionary behavioural outputs, like covering one's ears, moving the conversation to another location, or uttering "Wow, that noise is really bothersome." Note, however, that "access" in Block's sense does not have to manifest itself as a verbal report. "Reportability is a legacy of behaviorism that is less interesting than it has seemed. The more interesting issue in the vicinity is not the relation between the phenomenal and the reportable, but rather the relation between the phenomenal and the cognitively accessible" (Block 2007, 484). So, presumably, one could have access and still not be able to express this in any overt act of communication. Hence, "if Block is right, then there is not only a large part of our mental life that is inaccessible to us, namely the unconscious part; in addition, even a large part of our *conscious* mental life will then be inaccessible to us, since he argues that there are phenomenal yet inaccessible experiences" (Schlicht 2012, 310).

It is usually sound methodology to think that, if a posited object does not manifest itself in any overt way, the object in question does not in fact exist. Cryptozoology notwithstanding, a rare beast cannot be so rare that no one ever witnesses it. This, however, is the (purportedly hasty) reasoning that Block seeks to assail. Although one typically determines the absence of an aspect and/or faculty of consciousness by way of a contrast with its manifest presence in healthy persons, Block claims that such a move is inconclusive, since it does not fully exclude the possibility that the aspect and/or faculty in question might still lurk in an afflicted patient's mind. In fact, Block (1992) has explicitly accused functionalists of begging the question against phenomenal-consciousness (for a discussion of Block's stance, see Tye 1996, 291–295).

Exasperated by the elusive nature of these arguments, Daniel Dennett (1988) has claimed that talk of qualia is just a massive social delusion. We may all agree about what we are talking about, but there may be nothing more to this than our agreement. Block, by contrast, thinks that "it is obvious that P-consciousness is not a cultural construction" (1995a, 238; see also Block 1999). Although Block's distinction is still debated, most would agree with Chalmers that, at the very least, "[t]here is clearly a *conceptual* distinction here" (1997, 148). The question, then, is how best to handle this "conceptual" distinction. Addressing this, Güven Güzeldere writes:

> [C]ould it be that the particular way Block's distinction carves out phenomenal consciousness, separating it completely from its causal and functional aspects in accord with the "segregationist intuition," renders its investigation by means of scientific methods theoretically impossible? Put differently, could we be painting ourselves into a corner by a conceptual commitment to Block's distinction such that we end up with a number of straightforward problems about A-consciousness and a conjured-up "hard problem" of P-consciousness that in principle admits no solution? (Güzeldere 1997, 29)

Since too strong a reading of Block's distinction risks "segregating" the mind's qualitative dimension, I want to call on a conceptual distinction that is explicitly less than real yet more than nominal.

3.2 Prescission in Theory

The idea of representation figures prominently in philosophy of mind and the cognitive sciences, but there is a growing sense that it cannot do all the work we want it to. As Denise Gamble writes, "[a]n ontology of representations is a powerful tool for explaining some types of content. But not every internal stimulation or activation in mentality need be a representation. Is there no other conceptual framework for understanding phenomenology?" (1997, 150). There is indeed.

One of the central tenets of Peircean semiotics is that the very idea of representation, carefully unpacked, presupposes a three-place relation that cannot be sundered; that is, one which cannot be reduced to the dyadic or the monadic on pain of no longer representing (see Peirce 1998, 411). This does not, however, mean that qualia are representational. Semiotics does not say that a quale represents, but rather that representation involves a quale—there is no reciprocity (i.e., no monad is a triad).

Some (e.g., Colapietro 1989, 18) think that the term "representation" has suffered too much harm at the hands of sceptically-minded philosophers, but I am not prepared to give it up. Representations, from a semiotic point of view, are not a special class of objects such that certain (typically mental) things inherently have to represent while others can never do so. Much the opposite: the tone emphatically does not have to be the ground upon which interpretation pole-vaults to an object. If and when a tone has enabled semiosis (by being an icon), then there is no question of denying its all-important service. But, the whole point of prescission is that we can recognize quality as an ordinal First in such a relation, thereby incorporating into our theoretical picture the idea that a tone can stand for something else but need not do so. Thus, despite the unbreakable triadic bond which characterizes any representation, whatever is burdened with the logical duty of standing for something else—no matter what it may consist in—can be prescinded in such a way as to disregard its employment in that capacity.

This means that, following Block (1995b, 33–34), orgasms don't have to be "about" anything. But it also means that if one is led to infer from this "that something very pleasing is happening down *there*" (Tye 1995, 269), then, to that extent, the orgasm is acting as a sign (in this case, an index). From a semiotic standpoint, however, there is nothing about bodily feelings or sensations that make them more apt to serve as bearers of meaning, nor is there anything that bars a particular class from doing so. Block's talk of "mental paint" (1995b, 27–29; 2003), though couched in a mentalistic idiom, comes very close to the notion of tone. Even so, Block's proposal is less desirable because it implies (by its very name) that the issue of whether something is or is not a vehicle of representation—of whether it stands for something else to something—can somehow be answered by studying the nature of the candidate in question. According to the view I recommend, that is a misguided endeavour:

> [T]he *being* of the sign is the triadic relation itself, not the elements related or structured according to their respective roles [...]. The representative element within this triadic structure, which we loosely call a "sign," "in itself" is *not a sign at all*, but one of the three elements *necessary to the being of a sign*, one of the three legs on which the sign walks in working its way through the world, and, indeed, the "foremost" leg, insofar as it is the leg which takes the direct representative step in carrying a semiosis. (Deely 2005, 176, 178)

Block states that he does "not want to claim that there are non-representational phenomenal features of every experience or that when there are, these non-representational features form support [for] the representational features in the manner of a 'base'" (1995b, 28). That is exactly what Peircean semiotics claims, the tone being the ultimate ground one can reach. Having said this, one must keep in mind that "[w]hat is sign-vehicle one time can be significate [i.e., object] another time; and what is interpretant one time can be sign-vehicle the next time; and so on, in an unending spiral [...]" (Deely 2005, 178).

We cannot extract a quale and put it in a test-tube like we would a sample of red blood. We can certainly stare at the colour of blood, but considering that red tone apart from the token liquid requires a prescissive separation that no machine can (and will ever) provide. Prescission is "the act which our mind does when it separates by means of thought things that are in point of fact inseparable" (Foulquié and Saint-Jean 1962, 562; my translation). Block, in contrast, suggests that the fact that we can conceive of a quality not accessed in any overt state of consciousness is evidence that a distinct phenomenal-consciousness module truly exists. Here is how he describes qualia:

> [W]e have P-conscious states when we see, hear, smell, taste, and have pains [...]. Here is another reason to believe in P-consciousness without A-consciousness: Suppose that you are engaged in intense conversation when suddenly at noon you realize that right outside your window, there is—and has been for some time—a pneumatic drill digging up the street. You were aware of the noise all along, one might say, but only at noon are you *consciously aware* of it. That is, you were P-conscious of the noise all along, but at noon you are both P-conscious *and* A-conscious of it [...]. Only at noon is the content of your representation of the drill *poised* for use in rational control of action and speech [...]. The example shows the conceptual distinctness of P-consciousness from A-consciousness and it also puts the burden of proof on anyone who would argue that as a matter of empirical fact they come to the same thing. (Block 1997, 380, 386–387; in the original 1995 article, Block speaks of a "deafening" drill)

Stripped to its essentials, Block's argument can be summarized as follows: 1) Phenomenal-consciousness is conceivable without access-consciousness (i.e., the passage above). 2) Access-consciousness is conceivable without phenomenal-consciousness (i.e., the projected case of "super-blindsight"). Therefore: 3) We are entitled to distinguish phenomenal-consciousness from access-consciousness. The theorist working with the incomplete type/token distinction needs both premises to proceed to the conclusion—which is then glossed as proof that qualia exist as tokens. Indeed, Block has made his ontological commitments in this regard crystal clear: "Whether we use 'consciousness' or 'phenomenal consciousness,' 'awareness' or 'access-consciousness,' the point is that there are two different concepts of the phenomenon or phenomena of interest. We have to acknowledge the possibility in principle that *these two concepts pick out different phenomena*. Two vs. one: that is not a verbal issue" (2000, 133; emphasis added).

If we heed the insight that triadic relations can be decomposed without their involving a multiplicity of distinct objects, we can proceed straight to the conclusion after the first premise. Compare Block's remarks with the following passage by Peirce:

> Among phanerons [Peirce's name for phenomena] there are certain qualities of feeling, such as the color of magenta, the odor of attar, the sound of a railway whistle, the taste of quinine [...] I do not mean the sense of actually experiencing these feelings, whether primarily or in any memory or imagination. That is something that involves these qualities as an element of it. But I mean the qualities themselves which, in themselves, are mere maybes, not necessarily realized [...]. A quality of feeling can be imagined to be without any occurrence, as it seems to me. Its mere may-being gets along without any realization at all. [...] I suppose you will tell me that no such thing could be alone in the universe [...]. But I point out to you that *these things are only known to us by extraneous experience; none of them are either seen in the color, heard in the sound, or felt in the visceral sensation. Consequently, there can be no logical difficulty in supposing them to be absent*, and for my part, I encounter not the slightest psychological difficulty in doing so, either. (Peirce 1931–58, 1.304–305; emphasis added)

The similarity between Block's illustrations and Peirce's analysis is striking. With these two accounts now in plain sight, would it not be preferable to keep intact their common contention that a legitimate distinction is at play—while recognizing that it owes to our ability to "peel off" occurrences and glance in an abstract fashion at the qualities they presuppose? As I said in Chap. 2, my account of consciousness accepts a dual commitment but rejects a non-overlap. This lets one buy into Block's distinction without endorsing his claim of mutual independence.

The debt to Duns Scotus is apparent: existential inseparability indeed does not entail identity in definition. Of course, Scotus and Peirce were not the only philosophers to have grasped this. As Joseph Levine writes in discussing the problematic entanglements that accompany arguments for qualia: "One cannot infer from a variety of modes of access to a variety of facts being accessed" (1997, 546). Likewise, "[t]he identity theorist can admit a duality, or even a plurality, of different *types of knowledge* without thereby committing himself to a duality of *types of things known*" (Churchland 1988, 34). The fact that we can prescind a tone from a token is not a sufficient reason to think that a tone exists apart from its functional role(s). Semiotic theory can thus lend support to Block's insightful but embattled suggestions.

Prescission teaches us that underneath all the hubbub of thought, discourse, and that general "action of signs" which Peirce called semiosis, there is the tone: a monadic dimension that has the power to be the qualitative vehicle of representation but which in virtue of its ordinal primacy remains serenely ignorant of whether it is actually employed in so raucous an activity (see Peirce 1931–58, 1.422–426). By refusing to reify the features it prescinds, such an approach allows us to respect the distinct character of phenomenal experience without turning it into a chimera. It is not that access-consciousness and phenomenal-consciousness are tokens of different types; rather, the latter is the tone of the former's tokens.

This key claim (which cannot be expressed using the more limited vocabulary of types and tokens) explains why "A-consciousness and P-consciousness are almost always present or absent together" (Block 1995a, 242). Given that tones are not themselves occurrences, the steadfast accompaniment in no way means that the qualities at hand somehow "supervene" on their corresponding tokens (Chalmers 1996, 32–89; Kim 1990). Instead, the situation involves a species of subsumption.

Tim Bayne and David Chalmers write that "[t]he paradigm case of subsumption is the relation between a complex phenomenal state and a simpler state that is intuitively one of its 'components'" (2003, 40). I agree. But, whereas Bayne and Chalmers describe subsumption as "a relation among token phenomenal states" (2003, 40), my account takes the subsumption to extend farther, since it regards any token as subsuming a tone. Despite Bayne's endorsement of a mereological model (2010, 20–46), it never occurs to him to also sunder tokens. It can be done.

As we saw in Chap. 2, the type/token/tone distinction doesn't just names things differently, it counts things differently. Adopting the original Peircean trichotomy thus reshapes the debates. Chalmers summarizes Block's distinction in three points: "(1) one can imagine access without experience and vice versa; (2) access can be observed straightforwardly, whereas experience cannot; and, most important, (3) access consciousness seems clearly amenable to cognitive explanation, whereas phenomenal consciousness is quite perplexing in this regard" (Chalmers 1997, 148). To layer a summary of my own, the outlook I advocate gives good grounds to be wary of the symmetrical "vice versa" of (1), agrees with the gist of (2), and marshals semiotic tools which—when properly understood—allow (3) to appear less foreign from the standpoint of explicit understanding.

Centuries of semiotic inquiry have produced some hard-earned results which we can now distil to a handful of tenets (a store of achievements nicely summarized in the otherwise eclectic Nöth 1995, 79–80). Using the terminology employed by Armstrong (1989), philosophers of signs have learned that representations necessarily have to be "layer-cakes" (and cannot be unitary "blobs") on pain of no longer representing, and that this irreducible complexity in turn allows us to distinguish without extinguishing the constituents that make up a whole greater than its parts. Moreover—and this is of special importance to the "extended mind" conception in cognitive science (compare Clark and Chalmers 1998 with Skagestad 1999)—the previous holds true regardless of whether the triadic representation that pole-vaults on a qualitative vehicle to reach its object does so on the basis of a correlation that is subjective or objective, conventionally recognized or truly bound to its object.

Prescissive abstraction is a form of explanation that we can explain. Alluding to this subtle mode of distinction, Peirce wrote: "It may be noticed that, throughout this process, *introspection* is not resorted to. Nothing is assumed respecting the subjective elements of consciousness which cannot be securely inferred from the objective elements" (1992, 3–4; compare this with Heil 1988). The semiotic account of phenomenal qualia I have tendered would thus seem to meet the desideratum laid down by Dennett, who encouragingly stressed that "[t]he third-person approach is not antithetical to, or eager to ignore, the subjective nuances of experience; it simply insists on anchoring those subjective nuances to *something*—anything, really—that can be detected and confirmed in replicable experiments" (2001, 231). Let me now show how this perfectly sensible demand can be met.

3.3 Prescission in Practice

Block (1995a, 234) thinks that if one hears a nearby drilling noise and only later comes to notice it, there is a sense in which one was aware of the noise all along. On this view, one's phenomenal consciousness of the noise does not begin with one's noticing it. It would, of course, be difficult to tell how long the noise went on in one's head since, prior to becoming access-conscious, that qualitative experience was not involved in any detectable function(s). Were one to insist after the fact that one had heard it all along, one's insistence would come too late. The contentious philosophic issue concerns what, if anything, one's mental life was like apart from any explicit grasp.

Block crafts inventive thought-experiments to motivate his view, but he also calls on results obtained by the psychologist George Sperling. Sperling presented subjects with visual displays lasting 50 ms. The stimulus materials used by Sperling (1960, 3) typically included letters arranged into rows:

TDR
SRN
FZR

Subjects were presented with a blank screen immediately afterward. They were then asked how many letters they could identify (in the event of ignorance, subjects were instructed to venture a guess, so they always provided complete answers). Under these conditions, results showed a discrepancy between what subjects said they saw and what subjects could prove they saw. While subjects reported having seen all the letters, they could identify only a subset (usually a third or less) of these. Even so, subjects insisted on having been conscious, however briefly, of the whole visual display.

One quick explanation of this discrepancy between function and phenomenon is that subjects are simply confabulating when they report seeing all the letters. Recent experiments conducted by De Gardelle and his colleagues (2009) show that subjects will believe that letters were present in a display even when pseudo-letters are used. However, the question that interests me is whether, when genuine letters are used, subjects really did see them all—in spite of their inability to tangibly prove this. Block believes "that although one can distinctly see all or almost all of the 9–12 objects in an array, the processes that allow one to conceptualize and identify the specific shapes are limited by the capacity of 'working memory,' allowing reports of only about 4 of them" (2007, 487). Block's interpretation is controversial, because it states that subjects saw "all or almost all" the items shown, and moreover saw them "distinctly."

Commenting on Block's interpretation, David Papineau (2007, 521) thinks it is more parsimonious to see Sperling's results as motivating a distinction between an indistinct "scene" phenomenology and a more distinct "item" phenomenology. Presumably, only some items in an experienced scene receive cognitive attention. Indeed,

> Consciousness is the subject of many metaphors, and one of the most hardy perennials compares consciousness to a spotlight, illuminating certain mental goings-on, while leaving others to do their work in the dark. One way of elaborating the spotlight metaphor is this: mental events are loaded on to one end of a conveyer belt by the senses, and move with the belt—perhaps changing as they go—towards a fixed circle of light, which does not completely cover the width of the belt. Some mental goings-on fail to pass through the illumination, in which case they never become conscious. But others are illuminated, and thereby enter one's consciousness. Beyond the spotlight, at the other end of the conveyer belt, lies the filing cabinet of memory, into which some of the more garish or lurid of the belt's contents fall. (Byrne 1997, 103)

This metaphor informs a lot of work in neuroscience (see Crick 1993, 62). Looking at this generic picture of the mind, one might ask: what is the point of (or warrant for) countenancing unlit portions? Essentially, what Block is saying is that the portions that do not receive the spotlight of attention are nevertheless, in their own way, distinctly present to consciousness. At first blush, Block's stance seems unverifiable.

In keeping with Peirce's pragmatist maxim (1992, 132), I accept that, if "the object of our conception" does not "conceivably have practical bearings," then we have no basis to credit our concept with having an object. Or, to put that maxim in terms that speak directly to the current worries about consciousness: "Every form of thinking must betray itself in some form of expression or go undiscovered" (Peirce 1998, 18). The phenomenal-consciousness espoused by Block seems to violate this maxim, since it could never be detected. Yet, on reflection, I do not think Block's view necessarily fails to meet the demand for tangible effects. If we read the Peircean maxim carefully, it requires only that an object "conceivably" have practical bearings. So, unlike Jesse Prinz, who holds that "[a]vailability is not mere disposition" (2012, 105), I am willing to admit un-accessed contents, provided they support a power to be acted upon. Unnoticed drilling noises are thus admissible because, eventually, they are noticed.

Interestingly, by altering his initial experimental design, Sperling provided evidence that supports Block's claims:

> Sperling's clever idea was to test whether people really did see all or almost all of the characters and whether the phenomenology persists after the stimulus was turned off by playing a tone soon after the array was replaced by a blank. Subjects were to report the top row if the tone was high, the bottom row if the tone was low, and the middle row in case of an intermediate tone. The result was that subjects could report all or almost all the characters in any given row. (Block 2007, 487)

In the first round of tests, subjects had a bunch of letters flashed before them but performed poorly when asked to identify the items that they glimpsed. However, in a second round of testing, subjects were shown the same number of letters but were also given an auditory "retro-cue" that directed their mental focus to a given row. This allowed them to perform much better at the identification tasks. In both versions of the test, subjects reported seeing the whole visual display. However, it clearly became less risky (and even justified) to believe those introspective reports once the subjects made good on their claims in a practical way.

Subjects in Sperling's experiment were presented with a blank screen immediately after being shown the flashed image. Because the chemistry and physics of vision allow information to be available for longer than the strict emission of photons off a screen, impingement on sense organs can give rise to signs that persist once the causal exposure proper has ended. Block (2007, 487) quotes William James' *Principles of Psychology* on this lag time: "If we open our eyes instantaneously upon a scene, and then shroud them in complete darkness, it will be as if we saw the scene in ghostly light throught [*sic*] the dark screen" (James 2007, 645). Sperling readily granted the presence of such phenomenological afterimages. In his view, "[t]he question is not *whether* the observer continues to see the stimulus after the illumination is turned off, but for *how long* he continues to see the stimulus" (Sperling 1960, 20). Admittedly, a 50 ms stimulus does not last very long. Still, to the extent that an afterimage qualifies as an image, it can manifest enough determinacy to permit subsequent interpretations, enabling us to "read off details in it which were unnoticed whilst the eyes were open" (James 2007, 645).

The afterimage (partially) resembles the image initially flashed, so it counts as an iconic sign of that image. I will introduce iconicity more fully in the next chapter. For now, we need only point out that iconicity is at work in signs like diagrams. Diagrams are signs "which represent the relations […] of the parts of one thing by analogous relations in their own parts" (Peirce 1998, 273). For example, four suitably placed bottles of ketchup can show where four world leaders were sitting at a table (without suggesting that those leaders look like ketchup bottles). The grid of letters used by Sperling manifested such a diagrammatic organization. Given that there are fewer rows (or columns) than there are letters, the format allows information to be "chunked" (see Champagne 2016, 32–33) in a way that reduces one's cognitive and mnemonic load. This is why auditory cues were able to direct attention more efficiently. Importantly, these cognitive and mnemonic benefits occur regardless of whether one is looking at some "internal" mental image or some "external" worldly image. As Sperling nicely put it: "It is as logical or illogical to compute the information contained in a visual image […] as it is to compute the information in a visual stimulus" (1960, 21). Following Max Coltheart (1980, 184), Block urges us to distinguish between "neural," "phenomenological," and "informational" persistence. However, the shared quality that we find in iconicity skewers all these levels.

The expression "iconic memory" was introduced by Ulric Neisser (1967, 20) in the same book that coined the expression "cognitive psychology." However, neither Neisser nor his colleagues fully explored what is implied by this idea of storing and accessing information using an image-like quality. Philosophy of signs, which makes that exploration its chief business, recognizes icons as qualities that can be interpreted but do not have to be. This is exactly what Block needs, insofar as "Sperling's study firmly establishes that unreported letters *could* have been reported" (Prinz 2012, 103). I therefore believe that Block is depriving himself of powerful explanatory resources when he rejects the idea that "a 'visual icon' persists after the stimulus is turned off" (Block 2007, 487).

Block thinks that Sperling experimentally proved what his story about a drilling noise could only adumbrate. Strictly speaking, though, Sperling did no such thing. All that the subjects' correct performances can establish pertains to access-consciousness. Indeed, on the terms Block has set, the choice seems to be this: silently enjoy an experience—or report it and no longer deal with the phenomenal side of consciousness. This may seem like an unpalatable disjunction, but it follows directly from Block's (1995a) phenomenality/accessibility distinction. Accordingly, any inquiry into consciousness that manages to render that object of study tractable by some articulate method of detection risks being charged with avoiding the qualitative dimension of conscious experience.

Josh Weisberg notes that, in ordinary philosophical conversations, "'phenomenal consciousness' just means 'experience.' Many people have embraced this sense of the term and use it to roughly pick out conscious experience involving sensory quality" (2011, 438). In this "moderate" gloss, states besides a quality itself are allowed to enter the picture. By contrast, in what Weisberg calls the "zealous" gloss, the very presence of something besides the quality itself disqualifies the candidate from belonging to phenomenal consciousness. What-it's-like thus becomes "a monadic property [...] that a state has or lacks independently of its relations to other mental states" (Weisberg 2011, 439). Block's p-consciousness is zealous. Hence, "any explanation of phenomenal consciousness in exclusively cognitive, intentional or functional terms will fail to capture, without remainder, what is really distinctive about phenomenal consciousness" (Weisberg 2011, 438). On those terms, tampering with the data is inevitable. In an effort to sidestep this methodological challenge, Block has recently weakened his stance. Instead of saying, as he once did (Block 2007, 487), that phenomenal consciousness overflows accessibility, he now prefers to say that phenomenal consciousness overflows access. Looking at the Sperling experiments, his revised view "does not claim that any of the items in the array are cognitively inaccessible, but rather that necessarily most are unaccessed" (Block 2011, 567).

While a strategic shift from the "inaccessible" to the "unaccessed" blunts the force of many critics (like Cohen and Dennett 2011) who "think that a vote for overflow is a vote for inaccessible consciousness" (Block 2011, 574), I do not think a weakening of access succeeds in sidestepping the difficulties at hand. Block motivates his revised stance with an analogy: while only one lottery ticket wins, "this does not show that for any particular contestant the lottery is unwinnable" (Block 2011, 567). This point is well taken: an "unwon" ticket is not an "unwinnable" ticket. In fact, so long as a ticket could have won, the view espoused by Block connects well with the Peircean stance I want to defend. There is, however, an important disanalogy. Sticking with the lottery comparison, scientific observations work only with (and establish the existence of) winning tickets. Failure to be manifested in access thus makes an experiential ascription superfluous.

When a row-specific retrieval aid was added to Sperling's experimental design, subjects were able to access the previously unused portions of their visual experiences, thereby lending credence to their initial claim about having seen the whole scene. Of course, once third-person tasks corroborate first-person reports, it is easy to make confident pronouncements about what subjects were conscious of. Yet, what Block ostensibly fails to notice is that the very performances which made the second Sperling results interesting have to be supposed absent if they are to speak in an informative manner about what subjects experienced before they engaged in overt identifications. This makes the situation trickier. Block cannot open the fridge door and use that as evidence that the lights were on while the door was closed. I therefore argue that, if we want to follow Block and get philosophical mileage out of the Sperling results, we must employ prescission and suppose the tangible actions absent.

Prescission is what happens when, going against the facts, we suppose that some portion of a complex phenomenon is absent. This formal abstraction does not pretend to actually separate anything. Still, if we cannot permit it, then the very ingenuity which allows a cognitive scientist to study consciousness will always invite the retort that, "[s]ince any report relies on cognitive access, it cannot inform us about the presence of an inaccessible conscious representation" (Schlicht 2012, 319fn7). This is an abstract worry, and it can only be catered to by abstract means.

So, while many philosophers feel "a pressing need for a methodological approach that is capable of separating P[henomenal-consciousness] from A[ccess-consciousness] empirically" (Shea 2012, 308), I think that waiting for such an empirical separation is forlorn. It is pointless to try to wedge a "real" distinction between qualia and their practical effects. Indeed, "[t]here is a simple and fundamental reason why all attempts to get at the 'raw data' of experience fail: introspective evidence always arrives already interpreted" (Jack and Shallice 2001, 177). This observation accords with the Peircean semiotic claim that all intelligible cognitions are cloaked in a layer of Thirdness. As a result, there is simply no way to test what a pre-accessed state is like.

Even so, I believe that prescission can let us indirectly make sense of this idea. Let me therefore revisit Sperling's experiment and give it a new twist. My modified version will have three steps. The first step consists in looking at the same stimulus of letters that Sperling used. Sperling was more concerned with memory than I am, so in contrast with a 50 msec flash, my "slow-motion" version gives one ample time to view the display. The only directive one must abide by is to not act on what one experiences. One is instructed to stay still, like a Beefeater on duty.

I take it that, already, a subject following these instructions will have changed as a result of her causal exposure to the image. It may be possible not to act, but it is hardly possible not to react. However, the philosophic quarry is not reflexes, but "what it is like" for one to undergo an experience. Is there even such a qualitative dimension? Perhaps we can track a subject's involuntary eye movements (if any) to find out, but it would take an abductive leap to let those minute indices speak to the experience of a qualitative state. Extra support seems in order.

Step two of my slow-motion experiment thus requires a subject to perform a simple identification task: one must look at the rows and name the letters that one sees. Alas, the moment the subject breaks her silence and immobility to act on what she experiences, she makes herself amenable to functional description. We are now recording her linguistic outputs, not the intrinsic character of her experience(s). Step two can detect only access-consciousness.

This is where the third step of my slow-motion experiment comes in. Step one required subjects to look at the display of letters, silently, without acting on their experience. Step two required subjects to prove their experience of the whole visual display by identifying the various items. I take it that, under my slow-motion conditions, normal adults will have no trouble correctly naming all the letters. Having now identified all the letters, step three asks subjects to say in what order they named these items.

Note that, until we come to this third step, we have no evidence that subjects (privately or publically) represented their activity in step two, thinking to themselves "I am now reading from left to right." Step three of my modified Sperling experiment thus stands in the same relationship to step two that step two stood to step one. Indeed, the choice of word order has the same status that the unexpressed quality had: it was something that could have been verbalized or acted upon but was not. So, by parity, I argue that, if we are going to say that the visual experiences needed the identifications of step two in order to count as conscious, then we also have to say that the identifications of step two needed the meta-identification of step three in order to count as conscious. Surely this consequence is absurd. Hence, by Modus Tollens, I conclude that the visual experiences were consciously present prior to their overt identification.

Clearly, from a methodological standpoint, "[i]t is difficult to know exactly what is going on in the phenomenology of the subject who is undergoing the Sperling experiment, before being asked about the contents of a row" (Bayne and Chalmers 2003, 36). Block's (2011) claim that perceptual consciousness overflows cognitive access can thus look implausible. Yet, to my mind, it would be equally implausible to maintain that our experiences (visual or otherwise) come into being only once they are accessed. Ernest Sosa explores the possibility that "[o]ne's consciousness contains experiences that go unnoticed; unnoticed altogether, or at least unnoticed as experiences with an intrinsic, experiential character that they nevertheless do have" (2003, 276). I do not want to go that far. As I see it, a mental state must eventually leave some observable trace if we are ever to infer its presence. I thus have a lot of sympathy for the view that "a state's being conscious consists in one's being in some kind of higher-order mental state that represents that state" (Rosenthal 2005, 4). In fact, I see no reason why this process could not be open-ended (see Champagne 2009, 562–563). Even so, I do not think it follows that, without the experience that comes after it, a given experience is nothing.

This section set out to find "*something*—anything, really—that can be detected and confirmed in replicable experiments" (Dennett 2001, 231). Block maintains that

Sperling achieved this. But, in order to maintain this, Block has to violate the very distinction that he charges others with overlooking. Retro-cue or not, identification tasks track access-consciousness, not phenomenal-consciousness. Yet, presumably, even if Sperling had not devised ingenious ways of verifying what subjects saw, it still would have been true that subjects experienced the whole array of letters. Using a prescissive supposition of absence, I have tried to show how we can make sense of such a non-accessed experience. Qualitative experience is always tied to action, but qualitative experience can be prescinded from action.

3.4 The Stream and Still Pools of Consciousness

Consciousness has long been remarked for its stream-like complexity, one thought leading to another. In some disciplines, that stream may be the minimal unit of study (Dewey 1896). Semiotic inquiry, however, is neither phenomenology nor psychology, so no methodological constraint forbids the semiotician from adulterating this baseline of lived experience. In prescinding, we attend to some elements and deliberately neglect others (see Deledalle 2000, 5–6; Houser 2010, 95–96; Stjernfelt 2007, 246–255). In order to reach a lone tone like, say, the smell of a striploin steak, a lot of connections need to be supposed absent—including the knowledge that a striploin steak is responsible for the smell. Thus, in a way, "qualia are like the Cheshire cat's smile with the cat removed" (Fromm 2009, 261). But, as much as we might want to deride this removal, the fact is it can be done. It is incumbent upon us, then, to do justice to this unique (if admittedly bizarre) ability.

In reality, qualitative experiences are always caught up in actions, associations, significations, and recollections. William James was the first to describe this activity as a stream. But, the history of ideas seems to have forgotten that, in the same honest spirit, James also observed that,

> Like a bird's life, [the stream of our consciousness] seems to be made of an alternation of flights and perchings. The rhythm of language expresses this, where every thought is expressed in a sentence, and every sentence closed by a period. The resting-places are usually occupied by sensorial imaginations of some sort, whose peculiarity is that they can be held before the mind for an indefinite time, and contemplated without changing; the places of flight are filled with thoughts of relations, static or dynamic, that for the most part obtain between the matters contemplated in the periods of comparative rest. *Let us call the resting-places the 'substantive parts,' and the places of flight the 'transitive parts,' of the stream of thought.* It then appears that the main end of our thinking is at all times the attainment of some other substantive part than the one from which we have just been dislodged. (James 2007, 243; emphasis in original)

Peirce (1931–58, 8.89) considered this "one of the finest, if not the finest, passage" in James' *Principles of Psychology*. I agree. In my modified Sperling test, we have a perching when we just stare at the letters, silently. The action of signs gets spurred into flight once we name the letters that we see. We return to a perching, though, when we just read them, without additionally noticing what that action feels like. Flight resumes once we are instructed to notice and verbalize this feeling of reading.

Using the same terminology as James, David Rosenthal claims that "transitive consciousness can occur without intransitive state consciousness" (1997, 737). What Rosenthal says is true if it is analogous to the claim that one can drive on the freeway without stopping. However, I think that Rosenthal's claim is false if it is taken to mean that one can drive without retaining the ability to stop. As John Locke put it, it is not "any more necessary for the soul always to think, than for the body always to move" (1825, 54).

Essentially, I am arguing that the claim "An experience must be accessed" is not so much wrong as incomplete. Acts of access also have a feel. So, to be accurate, we should instead say "An experience must be accessed *in/by another experience*." This enlarged formulation is better, but it faces a regress that plays no part in our mental lives (Champagne 2009, 563). We can quickly halt this regress by switching from "must" to a modally weaker "can," thereby obtaining: "An experience can be accessed in/by another experience." This is the stance that I endorse. It captures well, I think, Peirce's account of semiosis:

> Peirce uniformly holds (1) that there is no such thing as a sign in isolation, every sign being a constituent of a sequential set of signs, so that apart from membership in this set, a thing has no meaning—or is *not* a sign; and (2) that in the sequential movement of signs thus ordered, the meaning of the earlier ones in the series is provided by or constituted by the later ones as their interpretants, until a conclusion (*logical* as a matter of course) is reached. Indeed, Peirce adheres so consistently to this view that he says, more than once, that signs, *as such*, form an infinite series, so that no conclusion of reasoning is forever final, being inherently open to having its meaning modified by further signs. (Dewey 1946, 88)

A request to name what one sees is a request to produce an interpretation. However, prescission allows us to suppose this interpretation absent. Doing so results in an experiential quality that plays no function. This quality could be considered "a nonsemiotic phenomenon, but to Peirce, it is nevertheless semiotic, since even if a sign refers only to itself it has the potential of producing an effect in a process of semiosis" (Nöth 2003, 14). Interpretation, however, is not incessant, so my account of consciousness is not a "process philosophy" (as in Williams 2016).

A subject presumably feels something when a visual stimulus is shown to her. Yet, in our quest to verify this, we forget that when a subject confirms what she saw (say, by correctly identifying it), she engages in an act which also has distinctive feel. We are, for the better part, unreflective creatures (Legg 2003). Whether or not one subsequently takes stock of what it feels like, there is clearly something "it is like" for one to read letters in the direction that one does. This qualitative character may become easier to scientifically track once it is accessed, but I have tried to show why it would be a mistake to make such access our philosophical starting point.

References

Armstrong DM (1989) Universals: an opinionated introduction. Westview Press, Boulder

Bayne T (2010) The unity of consciousness. Oxford University Press, Oxford

Bayne T, Chalmers DJ (2003) What is the unity of consciousness? In: Cleeremans A (ed) The unity of consciousness: binding, integration and dissociation. Oxford University Press, Oxford, pp 23–58

Block N (1978) Troubles with functionalism. In: Savage CW (ed) Minnesota studies in philosophy of science, vol 9. Minnesota University Press, Minneapolis, pp 261–325

Block N (1992) Begging the question against phenomenal consciousness. Behav Brain Sci 15(2):205–206

Block N (1995a) On a confusion about a function of consciousness. Behav Brain Sci 18(2):227–287

Block N (1995b) Mental paint and mental latex. In: Villanueva E (ed) Philosophical issues, vol 7. Ridgeview, Atascadero, pp 19–49

Block N (1997) On a confusion about a function of consciousness. In: Block N, Flanagan O, Güzeldere G (eds) The nature of consciousness: philosophical debates. MIT Press, Cambridge, pp 375–415

Block N (1999) Ridiculing social constructivism about phenomenal consciousness. Behav Brain Sci 22(1):199–201

Block N (2000) How not to find the neural correlate of consciousness. Intellectica 2(31):125–136

Block N (2003) Mental paint. In: Hahn M, Ramberg B (eds) Reflections and replies: essays on the philosophy of Tyler Burge. MIT Press, Cambridge, pp 165–200

Block N (2007) Consciousness, accessibility, and the mesh between psychology and neuroscience. Behav Brain Sci 30(5–6):481–548

Block N (2011) Perceptual consciousness overflows cognitive access. Trends Cogn Sci 15(12):567–575

Bornstein RF, Pittman TS (eds) (1992) Perception without awareness: cognitive, clinical, and social perspectives. Guilford Press, New York

Byrne A (1997) Some like it hot: consciousness and higher-order thoughts. Philos Stud 86(2):103–129

Chalmers DJ (1996) The conscious mind: in search of a fundamental theory. Oxford University Press, Oxford

Chalmers DJ (1997) Availability: the cognitive basis of experience. Behav Brain Sci 20(1):148–149

Champagne M (2009) Some semiotic constraints on metarepresentational accounts of consciousness. In: Deely JN, Sbrocchi LG (eds) Semiotics 2008. Legas Press, Toronto, pp 557–564

Champagne M (2016) Diagrams of the past: how timelines can aid the growth of historical knowledge. Cogn Semiot 9(1):11–44

Churchland PM (1988) Matter and consciousness. MIT Press, Cambridge

Clark A, Chalmers DJ (1998) The extended mind. Analysis 58(1):7–19

Cohen MA, Dennett DC (2011) Consciousness cannot be separated from function. Trends Cogn Sci 15(8):358–364

Colapietro VM (1989) Peirce's approach to the self: a semiotic perspective on human subjectivity. State University of New York Press, Albany

Coltheart M (1980) Iconic memory and visible persistence. Percept Psychophys 27(3):183–228

Crick F (1993) The astonishing hypothesis: the scientific search for the soul. Scribner's, New York

Davidson DH (1970) Mental events. In: Foster L, Swanson JW (eds) Experience and theory. University of Massachusetts Press, Amherst, pp 79–101

De Gardelle V, Sackur J, Kouider S (2009) Perceptual illusions in brief visual presentations. Conscious Cogn 18(3):569–577

Deely JN (2005) Defining the semiotic animal: how the postmodern understanding of human being supersedes the modern definition "res cogitans". In: Deely JN, Petrilli S, Ponzio A (eds) The semiotic animal. Legas Press, Toronto, pp 145–186

Deledalle G (2000) Charles S. Peirce's philosophy of signs: essays in comparative semiotics. Indiana University Press, Bloomington
Dennett DC (1988) Quining qualia. In: Marcel AJ, Bisiach E (eds) Consciousness in modern science. Oxford University Press, Oxford, pp 42–77
Dennett DC (2001) Are we explaining consciousness yet? Cognition 79(1–2):221–237
Dewey J (1896) The reflex arc concept in psychology. Psychol Rev 3(4):357–370
Dewey J (1946) Peirce's theory of linguistic signs, thought, and meaning. J Philos 43(4):85–95
Farrell BA (1950) Experience. Mind 59(234):170–198
Flanagan O (1992) Consciousness reconsidered. MIT Press, Cambridge
Fodor JA (1974) Special sciences (or: the disunity of science as a working hypothesis). Synthese 28(2):97–115
Foulquié P, Saint-Jean R (eds) (1962) Dictionnaire de la langue philosophique. Presses Universitaires de France, Paris
Fromm H (2009) The nature of being human: from environmentalism to consciousness. Johns Hopkins University Press, Baltimore
Gamble D (1997) P-consciousness presentation / A-consciousness representation. Behav Brain Sci 20(1):149–150
Gozzano S, Hill CS (eds) (2015) New perspectives on type identity: the mental and the physical. Cambridge University Press, Cambridge
Güzeldere G (1997) The many faces of consciousness: a field guide. In: Block N, Flanagan O, Güzeldere G (eds) The nature of consciousness: philosophical debates. MIT Press, Cambridge, pp 1–67
Heil J (1988) Privileged access. Mind 97(386):238–251
Houser N (2010) Peirce, phenomenology and semiotics. In: Cobley P (ed) The Routledge companion to semiotics. Routledge, London, pp 89–100
Humphrey N (1992) A history of the mind: evolution and the birth of consciousness. Simon and Schuster, New York
Jack AI, Shallice T (2001) Introspective physicalism as an approach to the science of consciousness. Cognition 79(1–2):161–196
James W (2007) The principles of psychology, vol 1. Cosimo, New York
Kim J (1990) Supervenience as a philosophical concept. Metaphilosophy 21(1–2):1–27
Kirk R (1974) Zombies versus materialists. Proc Aristot Soc 48:135–152
Kirk R (2005) Zombies and consciousness. Oxford University Press, Oxford
Legg C (2003) This is simply what I do. Philos Phenomenol Res 66(1):58–80
Levine J (1983) Materialism and qualia: the explanatory gap. Pac Philos Q 64(4):354–361
Levine J (1997) On leaving out what it's like. In: Block N, Flanagan O, Güzeldere G (eds) The nature of consciousness: philosophical debates. MIT Press, Cambridge, pp 543–555
Locke J (1825) An essay concerning human understanding. Thomas Davison, London
Milner AD, Rugg MD (eds) (1992) The neuropsychology of consciousness. Academic Press, New York
Neisser U (1967) Cognitive psychology. Appleton Century Crofts, New York
Nelkin N (1996) Consciousness and the origins of thought. Cambridge University Press, Cambridge
Nöth W (1995) Handbook of semiotics. Indiana University Press, Bloomington
Nöth W (2003) Crisis of representation? Semiotica 143(1–4):9–15
Papineau D (2007) Reuniting (scene) phenomenology with (scene) access. Behav Brain Sci 30(5–6):521
Peirce CS (1931–58) The collected papers of Charles Sanders Peirce. Harvard University Press, Cambridge
Peirce CS (1992) The essential Peirce: selected philosophical writings, vol 1. Indiana University Press, Bloomington
Peirce CS (1998) The essential Peirce: selected philosophical writings, vol 2. Indiana University Press, Bloomington

Pietarinen A-V (2006) Signs of logic: Peircean themes on the philosophy of language, games, and communication. Springer, Dordrecht
Place UT (1956) Is consciousness a brain process? Br J Psychol 47(1):44–50
Prinz JJ (2012) The conscious brain: how attention engenders experience. Oxford University Press, Oxford
Rosenthal DM (1997) A theory of consciousness. In: Block N, Flanagan O, Güzeldere G (eds) The nature of consciousness: philosophical debates. MIT Press, Cambridge, pp 729–753
Rosenthal DM (2005) Consciousness and mind. Oxford University Press, Oxford
Schacter DL (1989) On the relation between memory and consciousness: dissociable interactions and conscious experience. In: Roediger HL, Craik FIM (eds) Varieties of memory and consciousness: essays in honour of Endel Tulving. Erlbaum, Hillsdale, pp 355–389
Schlicht T (2012) Phenomenal consciousness, attention and accessibility. Phenomenol Cogn Sci 11(3):309–334
Searle JR (1992) The rediscovery of the mind. MIT Press, Cambridge
Shea N (2012) Methodological encounters with the phenomenal kind. Philos Phenomenol Res 84(2):307–344
Skagestad P (1999) Peirce's inkstand as an external embodiment of mind. Trans Charles S Peirce Soc 35(3):551–561
Sosa E (2003) Privileged access. In: Smith Q, Jokic A (eds) Consciousness: new philosophical perspectives. Oxford University Press, Oxford, pp 273–292
Sperling G (1960) The information available in brief visual presentations. Psychol Monogr Gen Appl 74(498):1–29
Stjernfelt F (2007) Diagrammatology: an investigation on the borderlines of phenomenology, ontology, and semiotics. Springer, Dordrecht
Stoerig P, Cowey A (1997) Blindsight in man and monkey. Brain 120(3):535–559
Tye M (1995) Blindsight, orgasm and representational overlap. Behav Brain Sci 18(2):268–269
Tye M (1996) The function of consciousness. Noûs 30(3):287–305
Weisberg J (2011) Abusing the notion of what-it's-likeness: a response to Block. Analysis 71(3):438–443
Weiskrantz L (1986) Blindsight: a case study and implications. Oxford University Press, Oxford
Williams J (2016) A process philosophy of signs. Edinburgh University Press, Edinburgh
Zahavi D (2007) Killing the straw man: Dennett and phenomenology. Phenomenol Cogn Sci 6(1):21–43

Chapter 4
Enlarging the Menu of Referential Options to Include Icons

Abstract In the previous chapter, I looked at why phenomenal-consciousness must, by definition, repel all experimental testing. In this chapter, I want to explore an important consequence of this, namely "the meaning objection." This objection asks how a qualitative experience with no detectable effects could ever be referred to by words or gestures. Bertrand Russell rightly argued that, without knowledge by acquaintance, linguistic descriptions can cohere yet fail to refer to anything in the world. Because similarity does not turn on proximity, I argue that a Peirce-inspired account of iconic reference can overcome this limitation. David Papineau has come very close to reconstructing such an account of iconicity, but I think that the lingering flaws in his account can be remedied by explicitly transitioning from Russellian philosophy of language to a more inclusive Peircean philosophy of signs.

4.1 The Indexical Phenomenal Concept Strategy

There are many arguments and thought-experiments which attempt to show, with various degrees of success, that consciousness has a qualitative dimension. One of the most memorable is Frank Jackson's "knowledge argument" (1982; 1986). Jackson invites us to consider a neuroscientist—named Mary—who, upon being raised in a strictly black and white setting since birth, is allowed for the first time to emerge from her isolated confines into a fully-coloured environment. Jackson's original claim is that, even if Mary had mastered a comprehensive functional account of colour prior to this exposure, her new experiences would give her an additional insight into "what it's like" to see a colour like red.

Using a terminology introduced by Bertrand Russell (1910–11), we can say that Mary in the cave can muster "descriptions" of colour vision but will lack a more intimate "acquaintance" until and unless she undergoes the experiences forbidden to her (Bigelow and Pargetter 2004). It is natural, then, to think that what Mary gains when she exits her confines is a new ability to indexically point to "this" quality (Perry 2001, 97, 146).

As promising as this account is, I believe it must fail. I agree that "[s]ome kinds of knowledge require distinctive forms of engagement between the knower and the known" (Bigelow and Pargetter 2004, 194). Judged by that standard, indexical

M. Champagne, *Consciousness and the Philosophy of Signs*, Studies in the History of Philosophy of Mind 19, https://doi.org/10.1007/978-3-319-73338-8_4

devices like "this" indeed bring us closer to their referents than linguistic descriptions can. Alas, "this" can pick out its target in space and time without any knowledge of "what" that target is, as evidenced by a locution like "I have no clue what this is." I therefore think that the standard roster of options is too coarse. According to the Peircean theory I defend, there are three ways in which one can know things, namely by convention, causality, and similarity.

Peirce offers us a non-psychological analysis of similarity and similarity recognition. "Though Peirce was [...] exceptionally knowledgeable in the prior semiotical traditions, his conception of the iconic sign [...] was developed neither by generalizing from cases and kinds of iconicity he had observed, nor by appropriating and developing a pre-existing theoretical conception of this sort of meaningfulness" (Ransdell 1986, 51). Using iconicity and the prescissive tools developed in the previous chapters, I want to argue that knower and known are at their closest when they share a quality, and that this is what would have to be involved in successful reference to a phenomenal state. In my account, qualia remain linguistically ineffable, but icons can nevertheless let one appreciate what "that" is.

My iconic account will thus contrast with versions of the "phenomenal concept strategy" that take indexicality to be central. The "phenomenal concept strategy" (Stoljar 2005) is an attempt to preserve physicalist commitments while accounting for why there appears to be a gap when it comes to explaining the qualitative dimension of consciousness (Levine 1983). Although there are several variants of this strategy currently vying for adoption (see Balog 2009), the general idea is that, since we have special concepts to pick out conscious states, whatever difficulties we have fitting consciousness into a naturalist picture may owe to the peculiar nature of those concepts. The strategy thus caters to those who accept that there is an epistemic gap between the material and the phenomenal but who deny an ontological gap (see Chalmers 1996, 165–168).

Indexicality is usually taken to be the direct mode of reference whereby language comes into contact with whatever it denotes. Indexicality is arguably "one of the best known features of Peirce's theory of signs" (Atkin 2005, 161). In the mainstream analytic literature, though, indexicality was introduced by John Perry (who did not discuss the historical lineage of the term). Perry's original intent was to challenge the (Fregean) view that propositions "have a truth-value in an absolute sense, as opposed to merely being true for a person or at a time" (1979, 6). The idea of indexing meanings to contextual circumstances is less controversial than the idea of indexing truth-values (compare Blome-Tillmann 2008 with MacFarlane 2014). In philosophy of language though, Perry's proposal has become a commonplace. We use some words like "this" in specific contexts, and these contexts fix what (in the world) our words point to (Kim 2010).

Some phenomenal concept strategists use this indexicality to explain problems regarding qualia. According to John O'Dea, a subject's particular spatial-temporal location explains the force of thought-experiments like the inverted spectrum (Shoemaker 1982). O'Dea argues, for instance, that a disagreement between an Earthling and a Martian about what each means by "I am in pain" would be "tantamount to a disagreement over whether Earth is *here* or Mars is *here*" (2002, 180).

O'Dea therefore surmises that "[t]he irreducibility of sensory terms [...] may be nothing more than a straightforward consequence of their indexicality" (ibid., 175).

This indexical account strikes a bargain with the skeptic. One can successfully refer to, say, the fact that one is now enjoying an experience of green, but the indexical sign that one uses to achieve this act of public reference cannot reach all the way to the quality of the experience (O'Dea 2002, 177). Since qualia are not captured by causality and since indexicality works precisely by exploiting causality, qualia are not captured by indexicality. This means that the qualitative dimension of consciousness cannot truly affect or be affected by discourse. On the further assumption—mistaken, as I hope to show—that indexicality is our ultimate means of reference, the privacy of qualia follows.

If we follow Russell (1910–11) and distinguish knowledge that comes from "acquaintance" and knowledge that comes from "description," we are left in a bind. Intrinsic qualities are not describable. By disjunctive syllogism, this ought to make those qualities knowable by acquaintance. Alas, acquaintance is not up to the task. Therefore, when prompted to convey the intrinsic character of a feeling, the convergence of two persons' verbal reports and/or behavioural responses remains inconclusive. People of course remain free to discuss how they feel, but they cannot *really* discuss how they feel. My goal is to eradicate this second clause.

To use an index, one has to place a sign-vehicle in the vicinity of its object. Such vicinity, however, should not be construed too literally. One can point to Alpha Centauri in the night sky; but one has to aim at a specific location if one wants to aid/elicit a specific interpretation. The fact that spatial coordinates matter in fixing the reference shows that, even if distance is not an issue, causal considerations are essential to explaining why/how anything can be non-arbitrarily "sensitive" to a context (see West 2012).

Upon emerging, Mary will need to use indices to convey the qualitative character of what she has just discovered. However, when Mary points to, say, a red rose and proclaims that "So *this* is what was meant by red," she cannot mean that particular flower, then and there. Were this what she meant, one could destroy the colour red once and for all simply by burning the flower. In an attempt to surmount this inability of indices to convey qualities, some have grafted the (much used but incomplete) type/token distinction onto indexicality to yield what Brian Loar (1997, 597) calls a "type-demonstrative." This is supposed to be a context-specific gesture and/or utterance that somehow manages to refer to "*That type* of sensation" (see Levin 2007, 88–89). I do not think this finessing succeeds. It makes perfect sense that one should be able to point to tokens, since these supply the presence needed for indices to do their referential business. But, if "type-demonstratives" were truly possible, one could literally see generality. Surely, one can see instances of a law, kind, or habit—but not the law, kind, or habit itself.

The ability to "see" types is supposedly achieved by "thick" perception (Masrour 2011). Of course, once an agent realizes that what she perceived was a token of a type, she can become convinced that she somehow "saw" the type. That, however, would be an embellishment of hindsight. Given that the stream of consciousness flows in a linear fashion, one way to test claims about so-called "thickness" would be to require a subject to ascertain—before any other tokens are experienced—whether there are in fact such other tokens. Clearly, a subject looking at a painting cannot tell, just by looking, whether it has ever been copied or mass produced. Rather than arguing that kind properties are only sometimes represented in experience, it seems more judicious to say that, when perceiving a single token, the most a subject is perceptually (and intellectually) entitled to answer is that a) it exists and b) another token like it *could* exist. Talk of "recognitional dispositions" (Siegel 2011, 100) captures this, but conflating the modal strength of (b) with the actuality of (a) would constitute a reification.

Loar asserts that "type-demonstratives" are "recognitional concepts" which, despite their recognitional status, "need involve no reference to a past instance," such that "[y]ou can forget particular instances and still judge 'another one of those'" (1997, 601). If one can do without past instances, what is the relatum in the judgement "another one of those"? That is a bit like saying that a sibling has no sibling(s).

These recent attempts to explain qualitative reference are influenced by other advancements in our understanding of reference. The standard analysis (from Aristotle to Kant to Frege) breaks "This gerbil" down into three components, insofar as a particular gets identified as a member or instance of a kind or universal by an act of judgement (see Peirce 1931–58, 1.485). It was a tangible advance of twentieth-century philosophy of language to stress that, irrespective of how one glosses the ontological status of universals or the epistemological workings of judgement, context of use would have to be involved in securing reference to a particular. Capitalizing on the well-deserved reputation of that account, "type-demonstratives" (and "thick perceptions") simply repeat this story to explain reference to (or perception of) universality. This implausibly outstretches the resources of indexicality. If one wants to refer to Gerbilhood by means of a situated sign-vehicle like "This," then, given the generality of the intended target, there should be no reason to prefer one particular gerbil over another. Yet, since in the end not all gerbils will be pointed to, it may rightfully be asked: why this one? The only sensible answer seems to be because it is in the vicinity of the utterer (needless to say, uttering "This" with no gerbils around would not accomplish much). The claim that demonstratives pick out tokens is therefore less contentious than the claim that demonstratives can somehow pick out types.

Russell held that "It is obvious [...] that we are acquainted with such universals as white, red, black, sweet, sour, loud, hard, etc., i.e., with qualities which are exemplified in sense-data" (1997b, 101). Taking Russell at his word, if one is in contact with an "exemplification," is it not a slide to construe this as contact with a universal? Russell added: "When we see a white patch, we are acquainted, in the first instance, with the particular patch; but by seeing many white patches, we easily learn to abstract the whiteness which they all have in common, and in learning to do this we are learning to be acquainted with whiteness" (ibid.). Again, if inductive generalization is needed to get at the targeted quality, how can this still count as acquaintance, which is defined as a "direct" mode of knowledge (Russell 1910–11, 108)? If one were truly capable of being acquainted with universals, these should simply present themselves to one, with no intervening particular(s). Needless to say, a subject-to-type access differs greatly from a (more plausible) subject-to-token-to-type access. Russell promises us the former but delivers only the latter (for a similar criticism, see Chalmers 2003, 233).

The challenges associated with pinpointing exact referents do not alter the fact that indexicality works by and on tokens, not types. It may not be obvious what to look for upon hearing "Look there!," but it is obvious that to find out one has to scan the nearby environment for a particular object or event. If, say, your friend points to a stranger who just walked into a crowded coffee shop, you might conceivably have some difficulty pinpointing who your friend intends; but that does not license you to roam the whole planet looking for the person she meant. There is plenty of room for failed attempts in my account of interpretation (Eco 1988). However, my account does not support the skeptical leap from fallibility to impotence (Champagne 2015b).

Picking out particular instances is by no means a negligible service, since it is part of what must happen if one is to grasp a likeness. Yet, if on full consideration we realize that thought, comparison, and other deliberate intellectual interventions are needed, then these need to figure in the official account. Merely pointing does not suffice.

One might reply that it is a matter of coming across the "right" exemplar. After all, if—in keeping with Peirce's account of abduction (Houser 2005)—the initial stage of establishing a sign-vehicle's referent is (and cannot help but be) a surmise, then that surmise could benefit from beginner's luck. If so, then the burden would be on the advocate of referential serendipity to explain why, in the vast majority of cases, we do not grasp types via a single token. In any event, confirmation that one indeed guessed a type right from the get-go can be revealed only by further action/experience, so one cannot "forget particular instances and still judge 'another one of those'" (Loar 1997, 601).

Ideally, a story of how one refers to the qualitative dimension of consciousness should be such that whatever post-emergence Mary does or uses to refer to her novel colour experience(s) is not something she could have done or used in her pre-emergence condition, otherwise Mary would not need to emerge. Symbols clearly do not live up to this demand, since prior to seeing red Mary can competently employ the word "red" found in her textbooks (the adjective "competently" is warranted because Mary can undoubtedly draw more red-based inferences than most lay persons). Therefore, with a twofold menu of symbols and indices, all hope must be placed on the latter option. Interestingly, the indices favoured by many phenomenal concept strategists do not fare any better than symbols. Indeed, if one were to ask pre-emergence Mary what she means by the word "red," she could very well point to a diagram of the appropriate wavelength and answer "This one now." Of course, we as outsiders are privy to the fact that Mary has brought the context-sensitive sign-vehicles "this" and "now" in the vicinity of an object ill-suited to truly convey what red "is like." But—and this is crucial—nothing in the indexical account permits us to regard her gesture as a blunder. I thus agree that "in order to be successful, the Phenomenal Concept Strategy needs [...] to explain how these concepts afford us a rich and substantial grasp of their referents" (Schroer 2010, 509–510). Icons furnish this substantial grasp.

In contrast with indices, icons work only if (and only because) the qualities match. What matters in iconicity is not that the sign-vehicle is near its object but rather that the sign-vehicle is like its object. Hence, if nothing in Mary's room is coloured, nothing in that room can be used to refer to colours. To be sure, the confines of pre-emergence Mary are filled with other icons. One pencil, for example, might resemble another pencil, and could thus be used to iconically refer to the other (and vice versa). Alas, familiarity with office supplies is not what is at stake, so emergence from the cave is needed for the relevant colour icons to become possible. Mary's eventual exit is therefore doubly enriching: not only does she get to experience something new, she also gains access to the various sign-vehicles capable of conveying the quality at hand. This is because, in iconicity, sign-vehicle and object are one and the same.

Iconic reference thus augments the "semantic" axis that is lacking in John Searle's (1980) "Chinese room" thought-experiment. Indeed, it should be noted that the distinction between semantics (vehicle-to-object), syntactics (vehicle-to-vehicle), and pragmatics (vehicle-to-interpreter) was introduced by Charles Morris in his influential 1938 International Encyclopedia of Unified Science paper on "Foundations of the Theory of Signs" (reprinted in Morris 1971, 13–71). Morris drew a methodological division of labour among those studying facets of a single object of study, semiosis (i.e., the full triad of vehicle-to-object-to-interpreter). One can *prescind* the axes discussed by Morris, but one can never divide them in fact.

In order to see how these semiotic ideas can advance current debates in philosophy of mind, we can take a closer look at the argument which led Jackson (2004a; 2004b) to abandon his previous conclusions. Howard Robinson (2008, 224) reconstructs Jackson's rationale as follows:

1. Reference to any *x* involves causal influence from *x* to the referential act.
2. If *x* is epiphenomenal then it has no causal influence on anything, so *a fortiori*, not on any referential act.

Therefore,

3. If *x* is epiphenomenal then it is something to which we cannot refer.

Therefore,

4. If qualia are epiphenomenal then they cannot be objects of reference.
5. Qualia (if they exist) are what we refer to by using our phenomenal concepts.

Therefore,

6. If qualia exist and are epiphenomenal then they can and cannot be objects of reference.

Therefore,

7. Epiphenomenalism about qualia is incoherent.

Jackson sides with the Russellian tradition in assuming that true descriptions must be reducible to acquaintances. As he puts it, "[o]ur knowledge of the sensory side of psychology has a causal source," such that when making claims all "our entitlement comes back to causal impacts of the right kinds" (Jackson 2004a, 418). I disagree. I contend that premise (1) is false, since there exists a mode of reference which, though not mind-dependent like symbols, does not rest on causality. Hence, I think the argument above is valid but unsound.

Premise (1) is pivotal to what has been called the "meaning objection" (Robinson 2012). Gilberto Gomes conveys this objection well: "But how can we refer to [our experience of red] if, by assumption, it cannot have any causal effect on our thought?" (2005, 78). My answer is: by means of an icon. Once we incorporate iconic reference in our overall picture, the terms of the debate shift: working out the logic, (3) and (4) become false, so (6) and (7) no longer follow. Let me therefore explore this referential option in greater detail.

4.2 Getting in Touch without Touching

Some philosophers of mind think that "[i]f qualia represent then it is plausible that they represent non-conceptually. That is, they do not have language-like structure but rather are akin to pictures [...]" (Balog 2009, 296). This is true, but the allusion to pictures can be limiting. A perfume, for example, is an icon, even though in resembling the smell of, say, lavender, it is in no way pictorial. Still, for better or for worse, the image has become a paradigmatic exemplar of iconicity. Prinz (2002, 25–32), for example, speaks of "imagism" in the cognitive sciences. Mental images have often been viewed with suspicion, but it is worth noting that even a critic like Zenon Pylyshyn is careful to stress that "the existence of the experience of images cannot be questioned. Imagery is a pervasive form of experience and is clearly of utmost importance to humans. We cannot speak of consciousness without, at the same time, implicating the existence of images" (1973, 2).

Peircean semiotics gives us tools to handle such phenomena. The virtue of focusing on the technical notion of icon is that it compels us to bear in mind that these signs are defined in virtue of the sort of referential relation they sustain: to bear an iconic relation is to guide interpretation by exploiting a qualitative bond that would exist regardless of whether another (similar) object or interpretation was present. I am applying this idea to debates in philosophy of mind, but there is nothing in the concept of iconicity *per se* which says that it must involve a similarity between a mind and a world. Since the similarity is between whatever plays the role of sign-vehicle and an object, iconicity can just as easily encompass world-world similarities. One might, for example, take "s" (on this page) to iconically stand for "s" (on this page). In fact, I think we should keep this sort of "sideways" confirmation as our gold standard when discussing more troublesome issues pertaining to consciousness.

As an icon, a quality can refer to that same quality. Here it is the shared tone—not the proximity of the tokens—that underwrites an interpretive passage. Peirce gave us tools to carefully track in what respect(s) something stands for something else to something. The triadic relation is what renders cognition possible, but the different parts of this relation are not always equally responsible for the conveyance of meaning. An index "is a sign which would, at once, lose the character which makes it a sign if its object were removed;" whereas an icon "is a sign which would possess the character which renders it significant, even though its object had no existence" (Peirce 1931–58, 2.304). Semiotics does not force one to choose between these different modes of reference. We can have both, at the same time, yet carefully prescind them.

Iconicity exploits the quality of Firstness, indexicality exploits the pairing of Secondness, and symbolicity exploits the arbitrary association of Thirdness. Hence, the philosopher of signs must always ask: in what respect is something being used as a sign? The answer to this question is constantly shifting. For instance, if a lime is used to signify a golf ball, then what matters is the shared spherical character. If the same lime is used to signify a blade of grass, then what matters is the shared green colour. If the lime is used to indicate a location, then what matters is its material occurrence in time and space. The taxonomies of philosophy of signs strive to clarify all this. Different supports constrain the sort of referential relation(s) that a sign can have. Conventional imputations must be re-applicable, so only as a type can a sign-vehicle have a symbolic bond to its object. Causation requires particulars, so only as a token can a sign-vehicle have an indexical bond to its object. Similarity requires a shared quality, so only as a tone can a sign-vehicle enjoy an iconic bond to the quality referred to.

We can use these semiotic distinctions to improve our understanding of conscious phenomena. Loar asserts that "[p]henomenal qualities vary in generality: I can note that a state of mine has what all smells share, or that it is the smell of new mown grass" (1990, 81). Viewed from the standpoint of philosophy of signs, Loar glosses over several distinctions. To say that a given quality is shared by other experiences is already to enter into some sort of comparison, and thus to take one quality as the (in this case, iconic) sign of another (or others). Also, to say that a smell is a smell of something besides that smell (like new mown grass) is to confess that the quality at hand has already entered into semiosis (Peirce 1998, 320). Now, there is nothing wrong or inherently problematic in noting a similitude between qualities, nor is there anything wrong or inherently problematic in the idea of taking a quality as the quality "of" a certain thing. What is wrong and problematic is the assumption that one can do all this whilst handling the simple quality itself, irrespective of its functional role or involvement in relations.

An icon is a sign-vehicle that has its representational power intrinsically, solely in virtue of the quality it has/is. A blue thing, for example, will retain an ability to relate to similar blue things, even if no other blue tokens actually exist. Its tone is a sufficient guarantor of a (realized or unrealized) iconic bond. Peirce drew on the views of Duns Scotus in crafting this account (Boler 1963). For Scotus, "this white thing can exist without similarity. If another white thing comes into being, then similarity begins to exist in this white thing. Hence, the foundation of the relation can exist without the relation" (Weinberg 1965, 101). This may be what Loar was trying to express with the claim that "[y]ou can forget particular instances and still judge 'another one of those'" (1997, 601). However, one must not gloss over the fact that, when a similar token has not entered the picture, the similarity of the lone tone is merely potential (and so cannot allow judgements like "another one of those").

Chalmers is right that Mary is able "to think demonstrative-qualitative thoughts in which both a demonstrative and a qualitative concept are deployed" (2004, 186). Prescissive analysis allows us to make sense of this double-duty. Consider the following situation:

> In reference to its object, this footprint is a perfect icon, although reversed like the image of a person looking at himself in a mirror. But it is *at the same time* the index of a presence on the island, and not just any presence [...]. The sign in itself has its own existence, an existence of a non-sign, one might say, just as an ambassador, although representing his country, is what he is in reference to himself [...]. (Deledalle 2000, 105)

If the footprint leads interpretation to a foot in virtue of its similarity with that foot, then it is the outline (of either the foot or the imprint) that matters. What permits iconicity is the quality of the sign-vehicle, not its causal impact with an object. If, however, a footprint leads interpretation to a foot in virtue of the causal contact it had with a foot, then it is the actual soil-foot dyad that matters. As for the word "footprint," nothing but interpretation holds its reference together. The symbol/index/icon distinction thus marks out three different ways sign-vehicles can be linked to their objects. Peirce rightly insisted that these referential relations "are all indispensable in all reasoning" (1931–58, 1.369; see his 1998, 10).

Recently though, it has become fashionable among some Peirce scholars to exclude similarity from the semiotic repertoire. In at least one instance, Peirce wrote that "a great distinguishing property of the icon is that by the direct observation of it other truths concerning its object can be discovered than those which suffice to determine its construction" (1931–58, 2.279). This asymmetry between construction and discovery certainly deserves further study, but it seems to have gained an uncritical acceptance. In fact, some papers have begun to slide from saying, as Peirce cautiously did, that the construction/discovery asymmetry is "a" distinguishing feature of iconicity, to saying that it is "the" distinguishing feature. Chiara Ambrosio, for instance, makes it look like Peirce put the construction/discovery asymmetry "at the centre of his notion of iconicity" (2014, 263). In so doing, she follows Christopher Hookway, who held that "[t]he key of iconicity is not perceived resemblance between the sign and what it signifies but rather the possibility of making new discoveries about the object of a sign through observing features of the sign itself" (2000, 102). Exegetically, this attempt to bypass similarity faces a tall hurdle since, as Chevalier (2015, 45) points out, Peirce used the term "icon" and "likeness" interchangeably. I do not want to veer into an exegetical dispute, so let me give tangible reasons to reject the construction/discovery view and retain the similarity view.

Suppose I use a smart phone to text you the following letter: M. Then, one minute later, I text you this letter: R. I instruct you to write all of these letters down as they come. Thus, keeping pace, you eventually receive and transcribe the following sequence: A, N, D, M, R, S, D, U, R, and so on. After a while, you realize that, unbeknownst to you, you are constructing the novel *Harry Potter and the Sorcerer's Stone*. Now, a book is a symbol, if anything is. Yet, the sequence of letters fully satisfies the definition of an icon suggested by the construction/discovery view, since by the direct observation of it other truths concerning its object can be discovered than those which suffice to determine its construction. You can learn about the exciting adventures of the students at Hogwarts, discover idiosyncratic tidbits about their personalities, and so on—just by assembling a list of typographical characters.

Consider this other counter-example. I have just said that a blue tone is an icon of any blue tone. Yet, taking the construction/discovery view at face-value, it may be asked: what "more" can one possibly learn from the observation of a simple blue tone? To say, for example, that blue is the colour of oceans, or that blue is a peaceful colour, is to import an external piece of information that is not itself present in that colour. As thinking agents with past experiences, we are privy to collateral data, so we can always let blue stand for more. Yet, strictly speaking, nothing more than blue can be learned by observing solely that lone quality. Hookway is clear that "a sign resembles its object if, and only if, study of the sign can yield new information about the object" (2000, 102). Although the similarity between blue and blue is arguably a paradigmatic exemplar of iconicity, it fails to fit this construction/discovery demand.

I submit that a tenable definition of iconicity ought to have excluded my first counter-example about the string of letters and included my second counter-example about the simple colour blue. What Peirce said about the potential for discovery might be applicable to diagrams, but it would be a mistake to abandon the similarity-based view of iconicity.

Philosophers have often been squeamish about countenancing similarities, so it is predictable that Peirceans trying to conform to mainstream expectations should jump at the opportunity to switch to a definition that spares them from endorsing what is, in some circles, an embarrassment. Despite helping herself to the term "icon" and purporting to develop "A General Theory of Signs," Ruth Millikan (1984, 83–158) does everything she can to avoid countenancing real similarities. Likewise, despite his interest in the varieties of reference, Gareth Evans refuses to acknowledge mind-independent similarities as a possible channel. Evans accepts that one thing could resemble another only "if it strikes people as like that other thing" (2002, 292). Analyses of similarity "anchored in the reactions they occasion in people" (Evans 2002, 294) have been amply explored—even by semioticians like Charles Morris (1971), Millikan's teacher. However, such behavioural approaches leave unanswered (or rather unasked) why these reactions occur in the first place.

Looking back, it is distressing to see how early iconicity got discarded, and how thin the grounds of that dismissal really were. One stock complaint is that pictures are misleading. "Of course, pictures may be misleading. But, so may sentences"

(Moktefi and Shin 2013, v). Another complaint, initially voiced by Nelson Goodman (1976, 38), is that everything is in some way similar to everything else. This, on the face of it, is untrue. Clearly, I am more like you than I am like a cloud of helium. Of course, one could add a host of stipulations to play up the similarities and narrow the difference(s), but then the stipulations would account for the (stretched) sense of sameness. In any event, even if it could be shown that everything can resemble everything, that would still be insufficient to discard similarities from the semiotic repertoire. After all, every material thing is (by transitivity) currently in some sort of causal interaction with everything else, but we hardly dismiss causality on that account (for a stepwise rebuttal of Goodman's grievances, see Stjernfelt 1999).

Frederik Stjernfelt contends that similarity is not a defining feature of iconicity, since "[s]imilarity is generally symmetrical: if *a* is similar to *b*, then *b* is also similar to *a*; while sign-relations are generally asymmetrical: if *a* signifies *b*, it does not follow that *b* signifies *a*" (2007, 49). It is true that similarity is symmetrical. It is also true that, in any triadic sign, interpretation will impose an asymmetrical sense of direction going from sign-vehicle to object. However, it is important to stress that this asymmetry is beholden to interpretation, not to the sign-vehicle or object. Since the overlaying of a means-end order on qualities that are otherwise identical can just as easily be turned the other way around, every object iconically referred to by a sign-vehicle is at the same time a potential sign-vehicle in the opposite direction. This is obscured by the twin facts that there is no reason to prefer one quality over the other, yet an interpretant will always privilege one sense of direction in a given instance. To the extent the sign-relation truly latched on to a real similarity between *a* and *b*, nothing in principle would have barred the reverse from happening, letting *b* do the "standing for." This means that, if an ordinary person looks like a famous movie star, then it is as legitimate to find that the famous movie star looks like the ordinary person (for each seesaw on this biconditional, a new interpretant is spawned).

Resemblance is mind-dependent in the sense that there must be an organism with an appropriate sensory system to deem one experience to be similar to another experience. The Peircean account that I promote has plenty of room for the effects which icons can have on such organisms. Still, my account regards those interpretants as effects, not causes, of underlying similarities. They come Third. For example, it is imperative to the evolutionary success of camouflage that the likeness between, say, an insect and a leaf, be truly mind-independent and not merely a wilful association (Maran 2003; Sebeok 1976, 1440–1441). Even in the absence of intention and causality, interpretation—which in semiotics is not the sole preserve of humans (Sebeok 2003)—can capitalize on the shared quality to let one thing stand for another.

Because an icon presupposes nothing but its own quality, its referential power "is not necessarily dependent upon its ever actually determining an Interpretant, nor even upon its actually having an Object" (Peirce 1998, 273). Consequently, the only way to eliminate the iconic potential of a tone is to eliminate that tone itself. Short of doing so, the ability to be linked to similar things always lies in wait, in germinal form, simply because any quality would resemble whatever would be like it. Of course, the mere talk of "another" tone would entail that we are no longer dealing with tones but with tokens, since juxtaposition or comparison presupposes numerical plurality. Still, when two tokens are related in virtue of their shared quality, it is the underlying tone they share that matters, not those particular tokens. Therefore, to understand iconicity, one has to prescind. Doing so reveals the icon to be an idle sign, something that "can only be a fragment of a completer sign" (Peirce 1998, 306).

The meaning objection rightfully asks "how can we refer to [our experience of red] if, by assumption, it cannot have any causal effect on our thought?" (Gomes 2005, 78). My suggestion has been that, because iconic signs ride on tones that are prior to numerically distinct tokens, they can get in touch with their objects without touching them.

4.3 How Low can we Go?

The Ancient Greek atomists were arguably the first to maintain that anything complex can be decomposed into simpler parts. Essentially, they drew on the following inference rule: P and Q, therefore P. The atomists deployed this inference in metaphysics, but Leibniz later saw that it is applicable to any domain, philosophy of mind included (Blamauer 2011). The logical implication of simplicity by complexity is so compelling that even an eliminativist like Paul Churchland must grant it:

> [T]he bulk of one's sensational life is characterized, not by simplicity, but by an extraordinary and ever-changing *complexity*. Listening to a conversation, looking around a flower garden, tasting a braised-lamb stew, smelling the aromas in a wood-working shop—our sensations in such cases display intricacies that are amazing. And not always obvious. A young child may not appreciate that the distinctive taste of her first ice-cream cone *resolves* itself into sensations of sweetness, creaminess, and strawberry. And it may take her awhile to learn that such decompositions are both common and useful to keep track of. For the complexities we encounter are indeed composed, quite often, of simpler elements or constituting dimensions. In time, we do learn many of those simpler dimensions. A dinner-table conversation contains my brother's unique voice as an identifiable element; the complex flower-garden displays the striking orange of a typical poppy blossom; the lamb stew displays the distinctive taste of thyme, sprinkled into the mix at the outset; and the smell of yellow cedar stands out from the other smells in the wood shop, at least to a seasoned carpenter. Each of these particular qualitative features of one's inner phenomenological life is certainly a *simpler* dimension of a more complex whole. (Churchland 2011, 32–33)

If we begin (as I believe we must) with a premise of complexity and grant (as I believe we should) that anything complex can be decomposed, then we are led to conclude that, in principle, such decomposition would have to bottom out at some point. So, in prescissively analyzing consciousness, how low can we go? With Peirce, I would say: as low as Firstness.

The first removals can be regarded as "real" in Scotus' sense. For example, "[w]hen I see the red book and hear the bird singing, there seems to be no good reason to deny that I could have a visually identical experience without hearing the bird singing, and so on" (Bayne and Chalmers 2003, 43). One way of separating these visual and auditory experiences is to close one's eyes or use ear plugs. However, if we want to decompose the complexity of conscious life into simpler and simpler parts, then eventually only a "formal" distinction will do.

In *The Philosophy of Logical Atomism*, Bertrand Russell wrote that "[t]he simplest imaginable facts are those which consist in the possession of a quality by some particular thing. Such facts, say, as 'This is white'" (1998, 59). Typically, when one says "This is white," one tries to identify/locate a white thing. Now, a white thing can be located only if it is amid non-white things. One of those non-white things must be the agent doing the pointing. Thus, pointing presupposes at least two relations: a relation to other things and a relation to an agent. This already involves a certain level of complexity. Hence, I do not think Russell was quite right when he said that "This is white" is the lowest one can go.

The sign-vehicle "This" is not white; in fact, here it is black (and, if spoken, it has no colour at all). Hence, in order to successfully use "This" as a sign of white things, one has to bring a token of "This" near a token white thing so that interpretation can relate the two. Russell may be right that "This is white" is the simplest *fact* imaginable, but I see no reason why factuality should be the stopping point of our analysis. As David Pears explains, Russell thought that "when we find that we cannot push the analysis of words any further, we can plant a flag recording the discovery of genuine logical atoms" (in Russell 1998, 5). When one does Peircean philosophy of signs, one can plant a flag further.

Peirce and Russell never met (Nubiola 1996, 283). Russell would later write that "I am—I confess to my shame—an illustration of the undue neglect from which Peirce has suffered in Europe" (foreword to Feibleman 1946, xv). Whatever their differences, Peirce and Russell both agreed that whatever is complex is composed of simples. "[N]o analysis," Peirce writes, "whether in logic, in chemistry, or in any other science, is satisfactory, unless it be thorough, that is, unless it separates the compound into components each entirely homogeneous in itself, and therefore free from the smallest admixture of any of the others" (1931–58, 4.548). Discussing qualia, Dennett remarks that "[s]ince they are 'simple' or 'homogeneous' there is nothing to get hold of when trying to describe such a property to one unacquainted with the particular instance in question" (1988, 385). Dennett finds this ineffability repugnant. The possibility that language cannot convey everything is, for him, unthinkable. I disagree (in fact, the belief that language cannot convey everything is arguably *the* defining belief of a philosopher of signs).

Because mainstream philosophy of language resorts only to indices, it will always have to contend with a minimal duality or Secondeness between pointer and pointed. Peirce's philosophy of signs countenances such indices, but it also collapses the duality of indexicality to reach an iconic merger. Unfortunately, with a few exceptions (like Giardino and Greenberg 2015), philosophy of mind is currently blind to this portion of the semiotic spectrum. Consider David Papineau. In the course of his investigations of consciousness, Papineau writes that "[a] first thought might be that perceptual concepts refer in virtue of the fact that exercises of them *resemble* their referents," and then immediately adds "I assume that this suggestion does not need to be taken seriously" (2002, 111). Why not? As an heir to Russell's

(1910–11) acquaintance/description distinction, Papineau assumes that our knowledge of the world derives from two (and only two) channels. Since I think Peirce is right to countenance similarity-based reference, I do not share Papineau's dismissal.

Ironically, no one working within a Russellian framework has come closer than Papineau to independently reconstructing an account of iconicity. Papineau qualifies as a phenomenal concept strategist, since he holds that the apparent distinctness of qualia is an artifact of the concepts we use to refer to them. Thus, like me, "Papineau diagnoses the apparent threats to physicalism posed by the phenomena of consciousness by locating the source of anti-physicalist intuitions in features of our thinking rather than in non-physical features of reality" (Crane 2005, 155).

According to Papineau, when a subject undergoes novel experiences, her "brain is lastingly altered in certain ways" (2003, 359). The alteration results in the acquisition of a "stored sensory template." Papineau uses this to explain what happens to Mary in the knowledge argument. The template that Mary stores can allow her "imaginatively to recreate and introspectively to reidentify an experience she could previously think about only in a third-person way" (ibid.). If Papineau is right that "the introspective identification of some experience requires that it is compared with some model or template stored in the brain," then "it would scarcely be surprising that we should need an original version of the experience in order to form the template for such comparisons" (ibid., 358–359). So far, I agree with all of this. Yet, Papineau stops short of acknowledging that the "comparison" at hand is similarity-based. This undermines his account. I therefore want to rid his account of the lingering Russellian assumptions that hold it back.

Using Russellian materials, Papineau (2002, 116–121) originally built what he called the "quotational-indexical" account of phenomenal concepts. However, he eventually came to think that indexicality imposes too strong a constraint on when and where phenomenal concepts can be exercised. Papineau's defection is reminiscent of Jackson's, as both were led to ponder the troublesome intersection of demonstrative reference and epiphenomenalism. The indexical sign relation at work in demonstratives turns on physical presence. One has to be suitably "near" something in order to refer to it by pointing. Likewise, something quoted must be present in order for the mentioning device to do its work. "Linguistic quotation marks, after all, are a species of demonstrative construction: a use of quotation marks will refer to *that word*, whatever it is, that happens to be made salient by being placed within the quotation marks" (Papineau 2007, 121). This will do in most circumstances. But, since qualia are not physically present in any straightforward manner, the analogy with quotation seems to bring little aid.

Led by these considerations, Papineau eventually rebuilt his account so that nothing turns on the actual presence of what is referred to. His recent work still retains the idea that phenomenal concepts involve the very quality referred to. This is the cornerstone of my account. However, the standard Russellian distinctions—specifically the type/token and description/acquaintance bipartitions—doom Papineau's efforts to failure.

Consider first the type/token bipartition. Clearly, any concept wedded solely to a particular token is bound to be severely limited in its use. After all, the taste of the ice cream one ate on the occasion of one's seventh birthday—if treated as a token—is a taste found in no other ice cream. The Peircean semiotician will of course notice that what is relevant in discussing the taste of ice cream is a tone; but a theorist unable to call on this crucial notion will recoil to her only remaining option when rejecting tokens as inappropriate. Predictably, then, Papineau wonders: "Can phenomenal concepts pick out experiential particulars as well as types?" (2007, 123). This question brings us back to the dead-end discussed earlier. Types cannot impact one's sensory organs, only tokens of types can; so any theory which hopes to account for phenomenal consciousness by invoking perceptual encounters with types is surely ill-fated.

> To be sure, we do say, as Peirce [1931–58, 4.537] pointed out, that there is but one word "the" in the English language. But this is no more to be taken *au pied de lettre* than is the statement that there is only one poisonous lizard in the continental United States [...]. There is not one *lizard* which is the "type-lizard," and many other lizards which are the token lizards. Likewise, there is not one *word* which is the type, and many other words which are the tokens. (Willard 1983, 287)

If philosophers of mind have been led by their recent discussion of phenomenal concepts to conclude that one can somehow "see" the type-lizard, then something has gone wrong along the way.

Using the full resources of the type/token/tone tripartition discussed in Chap. 3, one does not "reencounter" a type (Papineau 2007, 123); rather, a type is what permits one to encounter tokens of the same tone.

To further illustrate how a limited menu of options strong-arms Papineau into adopting unsatisfactory conclusions, consider the acquaintance/description bipartition. Knowledge by description can be detached from its worldly site of origin and communicated second-hand. Descriptions can therefore work just fine even though the object described is absent. Knowledge by acquaintance, by contrast, requires the actual presence of its object. In order to count as knowledge, *in absentia* discourse must eventually connect with *in praesentia* reference. Indeed, the whole point of Jacksons' knowledge argument is that mere description is insufficient; at least when it comes to conveying the content of experiential feels. Papineau accepts this, since by his lights "[i]t seems clear that the preemergence Mary does lack some concepts of color experiences" (2007, 111). Papineau is thus in a bind. Either he maintains that qualia can be referred to by description—which is what the knowledge argument denies; or he maintains that qualia can be referred to causally—which is difficult to make sense of in the case of qualia. Like Papineau, discussants who rely on Russellian notions bounce between these two options to no avail. I submit that, to dismount this seesaw, one needs to add the Peircean concept of iconic reference.

The employment of one yellow object to signify another yellow object by means of their yellowness (and not, say, their proximity to one another) does not permit one to dissociate questions pertaining to "the medium and the message" (to echo McLuhan), since these admit of a univocal answer, to wit, "yellow" (see Champagne 2015a, 537–539). As William Seager writes: "What can be called 'immediate consciousness' just has the peculiar reflexive property of allowing an appreciation of both the information being conveyed and the mode of conveyance" (1999, 93). Given this overlap, the quality which acts as a sign-vehicle cannot be omitted—on pain of omitting the very passage that makes that quality play a semiotic role in the first place. This explains why "[m]any phenomenal kinds can be referred to only through the content shared by experiences of the kind at issue" (Nida-Rümelin 2008, 310).

The icon is capable of "bringing its interpreter face to face with the very character signified" (Peirce 1998, 307). Echoing this, Papineau writes that "phenomenal concepts are too close to their referents for it to seem possible that those same concepts could refer to something else," since "the referent seems to be part of the concept itself" (2007, 132). Of course, given that an icon refers to a quality by being that very quality, this suggestion that the referent is "part of" the concept is not at all fanciful. Nonetheless, because Papineau lacks the notional resources to properly express this idea, he undermines his own conclusions. Tim Crane, for instance, complains that:

> [I]t seems to me entirely incredible that when one thinks about, say, pain, one must, as a necessary part of that very act of thinking, have an experience which *in any way* resembles pain. When the narrator of E. M. Forster's *Where Angels Fear to Tread* says that 'physical pain is almost too terrible to bear,' he is clearly intending to talk about pain in the phenomenal sense, pain as a feeling, an event in the stream of consciousness. In any normal sense of "phenomenal," then—any sense that relates it to its etymology and its traditional philosophical meaning—he is employing the phenomenal concept of pain. But in order to understand this remark, and therefore grasp the concepts which it expresses, I do not think I need to undergo, as a part of that very understanding, an experience which is in any sense painful. Yet this is what Papineau seems to be saying. (Crane 2005, 156)

Armed with the full symbol/index/icon taxonomy, we can come to Papineau's defence. All parties agree that describing something exclusively by symbols is a non-starter. Russell would be the first to agree. After all, "Russell was as aware as anyone else that not everything can be thought of by description, on pain of the whole system of identification failing to be tied down to a unique set of objects [...]" (Evans 2002, 45). Hence, tracking the inferential relations among words is not enough (Champagne 2016). The knowledge argument brings this out in a particularly memorable way. While sequestered in her cave from birth, Mary could have been taught by unscrupulous experimenters to take *Dungeons and Dragons* seriously and thereby make coherent functional responses about "ghouls" and "trolls." Hence, given that on one level "pain" is a symbol like any other, there is surely something right in Crane's claim that this lexical concept does not have to be painful. Yet, if one is to truly comprehend what that word refers to, then whatever quality one will have retained from such token episodes will itself be related to pain states by being able to indexically spot a similar state (if and when it presents itself) and have an iconic sense of "what" the quality is.

The second conjunct here is crucial, since it distinguishes the fine-grained appreciation of qualities that iconicity (via prescission) permits. The concept of pain can indexically refer to past experience(s), but at some point that concept will have to share the experiential quality itself, on pain of having no real clue what that feeling is like. In other words, if one asks the narrator of Forster's book "What is pain?," that narrator is free to answer "What I experienced last Friday;" and when asked what was experienced last Friday, he can in turn answer "What I experienced the Monday before"—and so on. I have no quarrel with any of this. Yet, if the person really possesses the relevant phenomenal concept, it cannot be anaphora all the way. Signs are not memes (Kilpinen 2008).

Peircean semiotics arranges icons, indices, and symbols in an ordinal fashion, such that the more developed grades of reference subsume the lesser ones but not vice versa (Peirce 1998, 9). Interestingly, this triadic pecking order is confirmed by empirical data. Consider the icon "IIIIII" and the symbol "6," which have a common referent but relate to it in different ways, the former non-conventionally, the latter conventionally. If symbolic reference could depart completely from iconicity, as Crane's criticism suggests, then one would expect the interpretation of Arabic numerals like "6" to be untainted by whatever cognitive and mnemonic limits plague its iconic counterpart "IIIIII." However, studies have shown that subjects asked to pick the largest among pairs of symbols like "4 versus 9" demonstrate a lag in their response times akin to figuring out "IIII versus IIIIIIIII." "These results strongly suggest that the process used in judgements of differences in magnitude between numerals is the same as, or analogous to, the process involved in judgements of inequality for physical continua" (Moyer and Landauer 1967, 1520; for more recent studies, see Carey 2009, 117–156).

Now, if there is evidence that a symbol such as "6"—a quantitative concept not exactly known for its poetic connotations—is in some way IIIIII-like, why should it be absurd to agree with Papineau that "[e]ven if imaginings of pains don't really hurt, they can share some of the phenomenal unpleasantness of real pains" (2002, 174)? One could also look to scientific and phenomenological studies which suggest that mere contemplation of a word or phrase primes the body for a host of motor and affective responses, such that reading "pain" is in some sense experiencing a trace of the relevant feeling (Gallagher 2006; Shapiro 2011, 70–113; Thompson 2007). So, if a novel would be written by a congenitally blind autistic literary savant who has no sense of fine grained emotion concepts, my Peircean semiotic theory predicts that, to the extent it would be considered a good novel by readers, this author would have relied on informant(s) who enjoyed the relevant experiences (these informants would then be the analogues of programmers who feed instructions and symbols in a Turing machine).

The interesting question is not whether concepts need to always resemble the things they refer to; there is a clear sense in which they do not (to that extent, Crane's criticism of Papineau is trivially right). The interesting question is whether mastery of symbols and indices alone could ever suffice to secure reference to the "feel" of experiences. After all, from a developmental perspective, iconic competence is often (and perhaps always) the gateway to symbolic competence. It is doubtful anyone ever mastered "6" without first mastering "IIIIII" (Resnik 1982, 98). Some biologists conversant with semiotic theory (e.g., Kull 2009; Deacon 1997) believe that this holds on the evolutionary ladder as well. Holistically drawing inferences along a symbol-to-symbol axis certainly remains possible (especially by machines that have never known otherwise). It is also possible for one not to feel a hint of pain when one reads or writes the word "pain." But, if one never does—anywhere, anytime, under any circumstance—one can hardly lay claim to what the word means.

Declan Smithies (2011, 22–25) argues that, unless a subject is phenomenally conscious of the object(s) of her demonstratives, she cannot rationally defend her claims when challenged. If challenged, it may be okay for Mary to refer to whatever her textbooks told her. But it is not okay for the authors of those textbooks to have never enjoyed the relevant icon(s). I thus agree with Peirce that "every indirect method of communicating an idea must depend for its establishment upon the use of an icon" (1931–58, 2.278).

This point can be put another way. Suppose that a subject were to possess a given colour concept solely in virtue of having been told about its relations to darker and lighter colours. Being told, say, that amethyst is midway between purple and pink could conceivably be informative to someone who has experienced purple and pink. If so, then that person's concept of amethyst would amount to a rule (involving several relata), and the unfamiliar quality sandwiched between purple and pink would become akin to a conclusion that can be inferred once one knows the relevant premises. Since the rule applies to a spectrum that is ordered, there is a temptation to dismiss the need to experience the midway quality itself (e.g., Churchland 1992, 102–110). Yet, that spectrum actually vindicates qualia. The colours sandwiching an unfamiliar shade remain unproblematic only provided that one does not slide the very sandwiching relation to either side. Without an iconic access to qualia though, one has to (constantly) make that slide. The premises adduced to secure a supposed inference of the quality therefore turn out to be insecure conclusions of their own, leading to a regress. This explains why "[o]ne cannot give an informative answer about seeing orange to the congenitally blind" (Pitt 2004, 31).

David Rosenthal believes that "[w]e cannot acquiesce in the unhelpful thought that we all know the qualitative state when we see it" (2005, 196). Why not? Rosenthal reasons, quite rightly, that if experiential familiarity with a quality is needed, then "[t]hat would amount to picking the phenomenon in purely ostensive terms, which leaves too much open for us to tell whether we can explain the phenomenon in a way that makes it intelligible" (ibid.). I differ from Rosenthal in being open to the possibility that, when we move past the range of symbolic description, we move past the range of intelligibility. Language is like a tour guide who can show you around Prague. No matter how good the guide, you still need to tour Prague.

On a common sense level, most of us grasp that when a dictionary defines a colour by citing other colours, its accomplishment is partial. The Jacksonian insight—present in Russell and developed by recent phenomenal concept strategists—is that symbols without indices are empty. Peircean semiotics takes this insight further by holding that indices without icons are empty. Papineau heeds the moral regarding the insufficiency of descriptions, adds to it a novel recognition of acquaintance's insufficiency with respect to qualities, and then tries to construct a model that could remedy this. The success of his positive suggestions is partial, but his desiderata are on target. In a coloured world, iconicity trumps indexicality as a more plausible way to explain reference to phenomenal experience(s).

4.4 Focusing on What Matters

I do not think that, before her exposure, Mary could use her exhaustive neurophysiological premises to draw a conclusion about the qualitative feel of a colour. Yet, even if we allow that Mary could somehow transmute her knowledge of quantities like wavelengths into a knowledge of colour tones (following the suggestion in Dennett 1991, 399–401), she would still have to pick out the right tokens to prove her mastery of colour concepts to her peers. Therefore, the question of reference does not go away (especially for the "heterophenomenology" championed by Dennett).

When Mary exits her black and white confines, the world provides her with the exemplars needed to prompt and convey the relevant experiences. So, suppose that, coming face to face with a red thing for the first time, she utters "This is what red looks like." On my analysis, her sentence involves symbolic, indexical, and iconic reference. The symbolic component is most obvious in the linguistic association that makes "red" stand for red. The indexical component, roughly captured by "this," tracks the things that prompt the qualitative experiences she wants to elicit in others (Chemero 2006, 64). These symbolic and indexical components are relatively well understood. However, the iconic component of Mary's utterance, roughly captured by "looks like," is far less understood. Even so, there is a growing sense that, without some sort of qualitative conveyance, Mary's demonstrative appeal will not fully succeed. I am trying to clarify this last step.

Let us assume for the sake of argument that the linguistic and gestural signs emitted by people are available to you, but that their minds are hidden. As Edmond Wright (2008) points out, mutual trust can quickly seal this divide. Instead of casting me as zombie for whom "all is dark inside" (Chalmers 1996, 96), you can decide to believe what I say when I report experiencing a colour. My prescissive analysis cannot eliminate this need for mutual trust, because focusing on a specific quality is a voluntary act. Of course, in order to verify whether convergence on a shared quality has been achieved, we have to start talking and gesturing again. Indeed, "when I describe this consciousness as the idea of a quality in itself, I merely, by the laws of speech, am forced to seize upon the character of separateness in order to let another person know what sort of consciousness I have in view" (Peirce 1931–58, 6.230). But, when the showing and telling is done right, subjects can genuinely appreciate what it's like for each other to enjoy a given phenomenal experience.

Seen in this light, symbols and indices are devices that fine-tune the coupling of two people's experiences. The better the fine-tuning, the less the epistemic risk. Papineau (2002, 66–67) doubts that an indexical construction like "This feeling" can select a specific quality in the manifold stream of consciousness. But, in a sufficiently impoverished setting, there is not much there to be wrong about. It would be gratuitous, for instance, for two people to stand in the vast white expanse of the science-fiction movie *THX 1138* and doubt that the other person is experiencing white too. The people in that setting would appreciate their shared experience by looking outward, not inward, so an account of qualitative experience does not necessarily have to depend on introspection (Delaney 1979).

Coming close to this account, Papineau writes that the "characteristic feature [of phenomenal thoughts] is that the conscious *referent* itself is involved in the *vehicle* of thought" (2006, 104). This unique feature of icons is brought out by the much-discussed transparency argument (Harman 1990; Tye 2002). Usually, in the transparency argument, one is told to pick a thing in one's surroundings and to "concentrate as hard as you can, *not* on the colours of the *objects*, but on the quality of your *experience* of those colours" (Carruthers 2000, 123). Yet, if one truly follows those instructions and "concentrates" on a colour, one can no longer contrast that quality with whatever other colour(s) delimit(s) it. So, unless one wishes to explicitly defend gestaltist commitments, the usual set-up is worded somewhat carelessly.

Despite changes in terminology, the terms of the current debate are essentially those captured in Searle (1983, 59): either a vehicle intervenes, in which case it blocks access to the object; or access to the object is achieved, in which case no vehicle intervened (Kind 2010). In my view, what the transparency argument establishes is that one would be impotent to tell whether the qualitative experience one undergoes is "internal" or "external" to one. Indeed, under the stringent exclusionary conditions just outlined, it would be just as reasonable to interpret a coloured expanse as an opaque screen as it is to interpret it as some physical object transparently present before one (Ransdell 1979). What deserves endorsement, then, is not one of these two glosses, but an agnostic mid-way, since both options are equally viable (until and unless further experience is allowed to enter the picture).

Take the example of a blue tone. As Peirce pointed out, there is nothing in this quality that mandates the existence of another quality. Speaking strictly as a logician, Peirce wrote that "[t]o suppose […] that there is a flow of time, or any degree of vividness, be it high or low, seems to me quite as uncalled for as to suppose that there is freedom of the press or a magnetic field" (1931–58, 1.305). Hence, strange as it may seem, "it is conceivable, or supposable, that the quality of blue should usurp the whole mind […]" (Peirce 1992, 290). Peirce insists that this supposition of absence can be performed around any quality. In what is perhaps the most radical move of his diagrammatic logic, Peirce removed the rim of a Euler circle (Deledalle 2000, 15). He explained his approach as follows: "We must begin by getting diagrammatic notions of signs from which we strip away, at first, all reference to the mind; and after we have made those ideas just as distinct as our notion of a prime number or of an oval line, we may then consider, if need be, what are the peculiar characteristics of a mental sign […]" (quoted in Colapietro 1989, 44). What, Peirce asked, remains of a Euler circle once its boundary is gone and its area spreads out without obstruction? The answer, Peirce held, is Firstness: a qualitative expanse that awaits further specification.

William Seager notes that "[o]ne of the core intuitions about intrinsic properties is that they are the properties that things have 'in themselves,' the properties that something would retain even if it was the only thing in the universe" (2006, 141). If we consider blue as "the only thing in the universe," it can hardly be called a "thing" anymore. That is why it is best characterized as a tone, not a token. It cannot be counted.

The blank sheet at the base of Peirce's diagrammatic logic is so basic that it applies to philosophy of mind. As James notes, when considering a white paper without any contrasts, there is "no 'pointing,' but rather an all-round embracing of the paper by the thought" (1975, 31). Firstness is the canvass of all communication and cognition. It is also the bare minimum that arguing parties must accept if their disagreement is ever to be resolved (see Peirce 1931–58, 4.431; as well as Pietarinen 2006, 60). Clearly, everything that is asserted can be asserted; everything that is thought can be thought. Thus, however bizarre it may seem, we have no basis for denying this modal pre-condition, because "[t]hat which underlies a phenomenon and determines it, thereby is, itself, in a measure, a phenomenon" (Peirce 1998, 2).

Since the unbounded expanse of Firstness is a ground that merely awaits sign-action, it cannot be communicated. Hence, if we want to be consistent, we have to concede that describing Firstness is, by definition, impossible:

> Stop to think of it, and it has flown! What the world was to Adam on the day he opened his eyes to it, before he had drawn any distinctions, or had become conscious of his own existence,—that is first, present, immediate, fresh, new, initiative, original, spontaneous, free, vivid, conscious, and evanescent. Only, remember that every description of it must be false to it. (Peirce 1992, 248)

Wittgenstein (2002, 187) suggested that some of our most problematic intuitions—and deepest mystical sentiments—stem from our ability to conceive of a bounded whole. Interestingly, Keltner and Haidt (2003, 303–304) have identified the two main traits of religious awe as vastness and the need to mentally accommodate that vastness. If we are dealing with a quality deprived of all relational contrasts, we get vastness, but the need for accommodation cannot be met. Peircean Firstness can thus inspire theological speculations (see Slater 2016). Strictly speaking though, no religious tenets follow from a commitment to Firstness. I may not be able to put into words the taste of pickled relish, but I do have to put a token spoonful of that condiment in a mouth to iconically convey what tone I mean. Hence, "[t]he ineffable is not something mystical or mysterious; it is merely that which evades description. It evades description, but it pervades experience" (Short 2006, 171).

Of course, every sheet of paper that I have ever encountered had a rim. So, judged by induction, what Peirce says about Firstness is false. Yet, if one goes in the opposite direction, by deduction, the claim that such complex facts can be prescissively decomposed seems to me undeniable. The subsumption of simplicity in complexity shows that, if an articulate musical symphony is possible, then so must be an "eternally sounding and unvarying railway whistle" (Peirce 1931–58, 1.305). However, because "[p]rescission is not a reciprocal process" (Peirce 1992, 3), one cannot infer a musical symphony from a simple sound.

Whereas Peircean iconicity engulfs the quality it refers to, Russellian acquaintance always keeps its object at bay. Peirce originated indices (see Atkin 2005; Sebeok 1990, 21), but Russell reworked the original notion, calling those signs "egocentric particulars" (Russell 1997a, 108–115). It was from Russell, not Peirce, that Perry got indexicality. Given the egocentricity, indices came to be seen as linguistic devices "about where one is, when it is, and who one is" (Perry 1979, 5). The monadic merger at play in iconicity was thus eclipsed by the assumption that such Russell-style demonstratives are "the mother and father of all information-based thoughts" (Evans 2002, 145).

A symptomatic statement can help us appreciate the scope of the missed opportunities. Chalmers matches my negative claim that what is involved in phenomenal knowledge is not an index. He writes: "Mary's thought involves attributing a certain substantive qualitative nature to an object that is identified demonstratively. The concept *R*—her qualitative concept of the sort of experience in question—is not a demonstrative concept at all [...]" (Chalmers 2004, 185). However, Chalmers does not match my positive claim that what is involved is an icon. The closest he comes to reference by shared quality is when he invents a "direct phenomenal concept" (Chalmers 2010, 267). Pausing to take stock of what he has gleaned from his deliberations, Chalmers writes: "All this is to say that there is something intrinsically epistemic about experience. To have an experience is automatically to stand in some sort of intimate epistemic relation to the experience [...]" (1996, 196–197). So far so good. Yet, Chalmers immediately adds: "[...]—a relation that we might call 'acquaintance'" (ibid., 197). Here, Chalmers takes a wrong turn. By using the Russellian label "acquaintance" and speaking of a "relation," Chalmers (2010, 283–291) inadvertently introduces a hiatus or gap between knower and known.

Galen Strawson takes the same wrong turn as Chalmers. Strawson comes close to iconicity when he observes that, in perceptual experience, "the having is the knowing" (2015, 219). Yet, Strawson immediately calls this "knowledge 'by acquaintance'" (ibid.). That is not what acquaintance is. Today's philosophers of mind want to capture a gap-free appreciation of qualitative content, but the original textual source could hardly be clearer: "I wish to preserve the dualism of subject and object in my terminology" (Russell 1910–11, 109). Paul Coates and Sam Coleman (2015, 5) are thus correct to identify the involvement of a two-term relation as a founding idea of the analytic tradition. This dualism has sometimes come under attack, but it has never been successfully replaced, so the notion of acquaintance resurfaces in the very attempts to correct it (see for example the "acquaintance relationalism" of Dorsch 2016). Eventually, one grows tired of the epicycles.

The point of switching to a Peircean paradigm is to explicitly make room for one-term relations from the start, under the rubric of Firstness. Strawson tries to achieve this with "presentation" (2015, 223). It has been tried before (see Gamble 1997). However, Russell made it clear that "the word acquaintance is designed to emphasize, more than the word presentation, the relational character of the fact with which we are concerned" (1910–11, 109). To say that A is "related to" B is to forever nudge B away from A, or at any rate ensure that A is never B (in the most demanding sense of "is"). Chalmers is simply working out a consequence of this when he asserts that "experiences are not red in the same sense in which apples are red. Phenomenal redness (a property of experiences or of subjects of experience) is a different property from external redness (a property of external objects)" (2010, 254). Although Chalmers thinks that "both are respectable properties in their own right" (2010, 254), his assumptions nevertheless leave him with two tokens to juxtapose: one "in here," the other "out there."

Presumably, one comes to know a token "out there" by means of a token "in here." Yet, no matter how alike those tokens are, the thin space of numerical distinctness between them is enough to cast doubt that one truly has referred to the quality in question. Chalmers is therefore right that a predicament like absent or inverted qualia "is occasionally found distasteful, but it is a natural consequence of the indexicality of the concept" employed to express phenomenal qualities (Chalmers 1996, 205).

Although indexicality has its place in Peircean semiotics, that framework also countenances an ideal case where what is signifying and what is signified are one and the same. The distinctions that let us appreciate qualia are prescissive, so I have no wish to deny that, "[w]hen Mary says, 'So *this* is what it is like!', what she refers to will almost certainly be a physical property of a physical event" (Bigelow and Pargetter 2006, 377). Such physical presence would suddenly matter if, say, Mary were to use a red rose to indicate to her colleagues that she passed by the laboratory while they were gone. Indices, however, do not exhaust the referential repertoire, so a red rose can fulfil roles that turn on its intrinsic quality. Deliberate focus on such a quality does not erase the fact that a token is needed to see a tone—particulars must impinge upon our senses. But, given that similarity is not proximity, it is misleading to claim in an unqualified way that "Reference to any x involves causal influence from x to the referential act" (Robinson 2008, 224).

Focusing on what matters (tone) does not always mean focusing on matter (tokens). This is why Peirce said that a quality "cannot actually act as a sign until it is embodied; but the embodiment has nothing to do with its character as a sign" (Peirce 1998, 291). In short, careful study of the conditions for the possibility of sign-action reveals a ground level where similarity becomes so pronounced that "[i]t is an affair of suchness only" (ibid., 163). When prescinding all the way to uncorrupted iconicity, we make ourselves unable to ascertain what is "inner" and what is "outer." This means that, contrary to the view expressed by Chalmers, if one looks solely at a quality, experiences are red in the same sense in which apples are red. Let me now explore the consequences of this.

References

Ambrosio C (2014) Iconic representations and representative practices. Int Stud Philos Sci 28(3):255–275

Atkin A (2005) Peirce on the index and indexical reference. Trans Charles S Peirce Soc 41(1):161–188

Balog K (2009) Phenomenal concepts. In: McLaughlin BP, Beckermann A, Walter S (eds) The Oxford handbook of philosophy of mind. Oxford University Press, Oxford, pp 292–312
Bayne T, Chalmers DJ (2003) What is the unity of consciousness? In: Cleeremans A (ed) The unity of consciousness: binding, integration and dissociation. Oxford University Press, Oxford, pp 23–58
Bigelow J, Pargetter R (2004) Acquaintance with qualia. In: Ludlow P, Nagasawa Y, Stoljar D (eds) There's something about Mary: essays on phenomenal consciousness and Frank Jackson's knowledge argument. MIT Press, Cambridge, pp 179–195
Bigelow J, Pargetter R (2006) Re-acquaintance with qualia. Australas J Philos 84(3):353–378
Blamauer M (2011) Is the panpsychist better off as an idealist? Some Leibnizian remarks on consciousness and composition. Eidos 15:48–75
Blome-Tillmann M (2008) The indexicality of "knowledge". Philos Stud 138(1):29–53
Boler JF (1963) Charles Peirce and scholastic realism: a study of Peirce's relation to John Duns Scotus. University of Washington Press, Seattle
Carey S (2009) The origin of concepts. Oxford University Press, Oxford
Carruthers P (2000) Phenomenal consciousness: a naturalistic theory. Cambridge University Press, Cambridge
Chalmers DJ (1996) The conscious mind: in search of a fundamental theory. Oxford University Press, Oxford
Chalmers DJ (2003) The content and epistemology of phenomenal belief. In: Smith Q, Jokic A (eds) Consciousness: new philosophical perspectives. Oxford University Press, Oxford, pp 220–272
Chalmers DJ (2004) Imagination, indexicality, and intentions. Philos Phenomenol Res 68(1):182–190
Chalmers DJ (2010) The character of consciousness. Oxford University Press, Oxford
Champagne M (2015a) A less simplistic metaphysics: Peirce's layered theory of meaning as a layered theory of being. Sign Syst Stud 43(4):523–552
Champagne M (2015b) Disjunctivism and the ethics of disbelief. Philos Pap 44(2):139–163
Champagne M (2016) Tracking inferences is not enough: the given as tie-breaker. Logos and Episteme 7(2):129–135
Chemero A (2006) Information and direct perception: a new approach. In: Farias P, Queiroz J (eds) Advanced issues on cognitive science and semiotics. Shaker Verlag, Aachen, pp 59–72
Chevalier J-M (2015) The problem of resemblance in Peirce's semiotics and philosophy. Versus 120:45–59
Churchland PM (1992) A neurocomputational perspective: the nature of mind and the structure of science. MIT Press, Cambridge
Churchland PM (2011) Consciousness and the introspection of "qualitative simples". Eidos 15:12–47
Coates P, Coleman S (2015) The nature of phenomenal qualities. In: Coates P, Coleman S (eds) Phenomenal qualities: sense, perception, and consciousness. Oxford University Press, Oxford, pp 1–30
Colapietro VM (1989) Peirce's approach to the self: a semiotic perspective on human subjectivity. State University of New York Press, Albany
Crane T (2005) Papineau on phenomenal concepts. Philos Phenomenol Res 71(1):155–162
Deacon TW (1997) The symbolic species: the co-evolution of language and the brain. W. W. Norton and Company, New York
Delaney CF (1979) Peirce's account of mental activity. Synthese 41(1):25–36
Deledalle G (2000) Charles S. Peirce's philosophy of signs: essays in comparative semiotics. Indiana University Press, Bloomington
Dennett DC (1988) Quining qualia. In: Marcel AJ, Bisiach E (eds) Consciousness in modern science. Oxford University Press, Oxford, pp 42–77
Dennett DC (1991) Consciousness explained. Little, Brown and Company, New York
Dorsch F (2016) Perceptual acquaintance and the seeming relationality of hallucinations. J Conscious Stud 23(7–8):23–64

Eco U (1988) On truth: a fiction. In: Eco U, Santambrogio M, Violi P (eds) Meaning and mental representations. Indiana University Press, Bloomington, pp 41–59
Evans G (2002) The varieties of reference. Clarendon Press, Oxford
Feibleman JK (1946) An introduction to Peirce's philosophy interpreted as a system. Harper and Brothers, New York
Gallagher S (2006) How the body shapes the mind. Clarendon Press, Oxford
Gamble D (1997) P-consciousness presentation / A-consciousness representation. Behav Brain Sci 20(1):149–150
Giardino V, Greenberg G (2015) Varieties of iconicity. Rev Philos Psychol 6(1):1–25
Gomes G (2005) Is consciousness epiphenomenal? J Conscious Stud 12(12):77–79
Goodman N (1976) Languages of art: an approach to a theory of symbols. Hackett, Indianapolis
Harman G (1990) The intrinsic quality of experience. In: Tomberlin JE (ed) Philosophical perspectives, vol vol 4. Ridgeview, Atascadero, pp 31–52
Hookway C (2000) Truth, rationality and pragmatism. Oxford University Press, Oxford
Houser N (2005) The scent of truth. Semiotica 153(1–4):455–466
Jackson F (1982) Epiphenomenal qualia. Philos Q 32(127):127–136
Jackson F (1986) What Mary didn't know. J Philos 83(5):291–295
Jackson F (2004a) Postscript on qualia. In: Ludlow P, Nagasawa Y, Stoljar D (eds) There's something about Mary: essays on phenomenal consciousness and Frank Jackson's knowledge argument. MIT Press, Cambridge, pp 417–420
Jackson F (2004b) Mind and illusion. In: Ludlow P, Nagasawa Y, Stoljar D (eds) There's something about Mary: essays on phenomenal consciousness and Frank Jackson's knowledge argument. MIT Press, Cambridge, pp 421–442
James W (1975) The meaning of truth: a sequel to "pragmatism". Longmans, Green, and Company, New York
Keltner D, Haidt J (2003) Approaching awe, a moral, spiritual, and aesthetic emotion. Cognit Emot 17(2):297–314
Kilpinen E (2008) Memes versus signs: on the use of meaning concepts about nature and culture. Semiotica 171(1–4):215–237
Kim H (2010) Two notions of indexicality. Semiotica 180(1–4):47–67
Kind A (2010) Transparency and representationalist theories of consciousness. Philos Compass 5(10):902–913
Kull K (2009) Vegetative, animal, and cultural semiosis: the semiotic threshold zones. Cogn Semiot 4:8–27
Levin J (2007) What is a phenomenal concept? In: Alter T, Walter S (eds) Phenomenal concepts and phenomenal knowledge: new essays on consciousness and physicalism. Oxford University Press, Oxford, pp 87–110
Levine J (1983) Materialism and qualia: the explanatory gap. Pac Philos Q 64(4):354–361
Loar B (1990) Phenomenal states. In: Tomberlin JE (ed) Philosophical perspectives, vol 4. Ridgeview, Atascadero, pp 81–108
Loar B (1997) Phenomenal states. In: Block N, Flanagan O, Güzeldere G (eds) The nature of consciousness: philosophical debates. MIT Press, Cambridge, pp 597–616
MacFarlane J (2014) Assessment sensitivity: relative truth and its applications. Oxford University Press, Oxford
Maran T (2003) Mimesis as a phenomenon of semiotic communication. Sign Syst Stud 31(1):191–215
Masrour F (2011) Is perceptual phenomenology thin? Philos Phenomenol Res 83(2):366–397
Millikan RG (1984) Language, thought, and other biological categories: new foundations for realism. MIT Press, Cambridge
Moktefi A, Shin S-J (2013) Preface. In: Moktefi A, Shin S-J (eds) Visual reasoning with diagrams. Birkhäuser, Heidelberg, pp v–xiv
Morris CW (1971) Writings on the general theory of signs. Mouton de Gruyter, The Hague
Moyer RS, Landauer TK (1967) Time required for judgements of numerical inequality. Nature 215(109):1519–1520

Nida-Rümelin M (2008) Phenomenal character and the transparency of experience. In: Wright E (ed) The case for qualia. MIT Press, Cambridge, pp 309–324
Nubiola J (1996) Scholarship on the relations between Ludwig Wittgenstein and Charles S. Peirce. In: Angelelli I, Cerezo M (eds) Studies on the history of logic. Walter de Gruyter, Berlin, pp 281–294
O'Dea J (2002) The indexical nature of sensory concepts. Philos Pap 31(2):169–181
Papineau D (2002) Thinking about consciousness. Oxford University Press, Oxford
Papineau D (2003) Theories of consciousness. In: Smith Q, Jokic A (eds) Consciousness: new philosophical perspectives. Oxford University Press, Oxford, pp 353–383
Papineau D (2006) Comments on Galen Strawson. In: Strawson G, Freeman A (eds) Consciousness and its place in nature: does physicalism entail panpsychism? Imprint Academic, Exeter, pp 100–109
Papineau D (2007) Phenomenal and perceptual concepts. In: Alter T, Walter S (eds) Phenomenal concepts and phenomenal knowledge: new essays on consciousness and physicalism. Oxford University Press, Oxford, pp 111–144
Peirce CS (1931–58) The collected papers of Charles Sanders Peirce. Harvard University Press, Cambridge
Peirce CS (1992) The essential Peirce: selected philosophical writings, vol 1. Indiana University Press, Bloomington
Peirce CS (1998) The essential Peirce: selected philosophical writings, vol 2. Indiana University Press, Bloomington
Perry J (1979) The problem of the essential indexical. Noûs 13(1):3–21
Perry J (2001) Knowledge, possibility, and consciousness. MIT Press, Cambridge
Pietarinen A-V (2006) Signs of logic: Peircean themes on the philosophy of language, games, and communication. Springer, Dordrecht
Pitt D (2004) The phenomenology of cognition or "what is it like to think that P?". Philos Phenomenol Res 69(1):1–36
Prinz JJ (2002) Furnishing the mind: concepts and their perceptual basis. MIT Press, Cambridge
Pylyshyn ZW (1973) What the mind's eye tells the mind's brain: a critique of mental imagery. Psychol Bull 80(1):1–24
Ransdell JM (1979) The epistemic function of iconicity in perception. In: Ketner KL, Ransdell JM (eds) Studies in Peirce's semiotic. Institute for studies in pragmaticism, Lubbock, pp 51–66
Ransdell JM (1986) On Peirce's conception of the iconic sign. In: Bouissac P, Herzfeld M, Posner R (eds) Iconicity: essays on the nature of culture. Stauffenburg Verlag, Tübingen, pp 51–74
Resnik MD (1982) Mathematics as a science of patterns: epistemology. Noûs 16(1):95–105
Robinson H (2008) Why Frank should not have jilted Mary. In: Wright E (ed) The case for qualia. MIT Press, Cambridge, pp 223–245
Robinson WS (2012) Phenomenal realist physicalism implies coherency of epiphenomenalist meaning. J Conscious Stud 19(3–4):145–163
Rosenthal DM (2005) Consciousness and mind. Oxford University Press, Oxford
Russell B (1910–11) Knowledge by acquaintance and knowledge by description. Proc Aristot Soc 11:108–128
Russell B (1997a) An inquiry into meaning and truth. Routledge, London
Russell B (1997b) The problems of philosophy. Oxford University Press, Oxford
Russell B (1998) The philosophy of logical atomism. Open Court, Chicago
Schroer R (2010) Where's the beef? Phenomenal concepts as both demonstrative and substantial. Australas J Philos 88(3):505–522
Seager WE (1999) Theories of consciousness: an introduction and assessment. Routledge, London
Seager WE (2006) The "intrinsic nature" argument for panpsychism. J Conscious Stud 13(10–11):129–145
Searle JR (1980) Minds, brains, and programs. Behav Brain Sci 3(3):417–424
Searle JR (1983) Intentionality: an essay in the philosophy of mind. Cambridge University Press, Cambridge

Sebeok TA (1976) Iconicity. Mod Lang Notes 91(6):1427–1456
Sebeok TA (1990) Indexicality. Am J Semiot 7(4):7–28
Sebeok TA (2003) Nonverbal communication. In: Perron P, Danesi M (eds) Classic readings in semiotics. Legas Press, Toronto, pp 173–182
Shapiro L (2011) Embodied cognition. Routledge, London
Shoemaker S (1982) The inverted spectrum. J Philos 79(7):357–381
Short TL (2006) Response to Carl Hausman. Semio Inq 26(2–3):157–176
Siegel S (2011) The contents of visual experience. Oxford University Press, Oxford
Slater G (2016) C. S. Peirce and the nested continua model of religious interpretation. Oxford University Press, Oxford
Smithies D (2011) What is the role of consciousness in demonstrative thought? J Philos 108(1):5–34
Stjernfelt F (1999) How to learn more: an apology for a strong concept of iconicity. In: Johansson TD, Skov M, Brogaard B (eds) Iconicity: a fundamental problem in semiotics. Nordic Summer University Press, Aarhus, pp 21–58
Stjernfelt F (2007) Diagrammatology: an investigation on the borderlines of phenomenology, ontology, and semiotics. Springer, Dordrecht
Stoljar D (2005) Physicalism and phenomenal concepts. Mind Lang 20(5):469–494
Strawson G (2015) Real direct realism: reflections on perception. In: Coates P, Coleman S (eds) Phenomenal qualities: sense, perception, and consciousness. Oxford University Press, Oxford, pp 214–253
Thompson E (2007) Mind in life: biology, phenomenology, and the sciences of mind. Harvard University Press, Cambridge
Tye M (2002) Representationalism and the transparency of experience. Noûs 36(1):137–151
Weinberg JR (1965) Abstraction, relation, and induction: three essays in the history of thought. University of Wisconsin Press, Madison
West DE (2012) The semiosis of indexical use: from degenerate to genuine. Am J Semiot 28(3–4):301–324
Willard D (1983) Why semantic ascent fails. Metaphilosophy 14(3–4):276–290
Wittgenstein L (2002) Tractatus logico-philosophicus (trans: Ogden CK, Ramsey FP). Routledge, London
Wright E (2008) Why transparency is unethical. In: Wright E (ed) The case for qualia. MIT Press, Cambridge, pp 341–366

Chapter 5
Seeing Things as They Are

Abstract In the previous chapter, I added similarity-based signs or icons to the standard menu of referential options. In this chapter, I want to explore the ramifications of this addition for perception. Peirce saw good reason to push his prescissive analysis of iconicity down to a single quality. I thus contrast his account with that of John Poinsot, a medieval philosopher who held that a sign always retains a slight distinction between the object it signifies and the vehicle that does the signifying. Because iconic merger gives us a way to vindicate ordinary perception, I favour Peirce's stance. The resulting view goes against the claim, put forward by John Locke, that the "secondary" qualities we experience are mere figments. I therefore do my best to undermine that Lockean worldview. Although some philosophers believe that putting qualities back into the world results in panpsychism, I try to find a less far-fetched way to express this.

5.1 Peirce's Merger Versus Poinsot's Buffer

A sign is composed of a sign-vehicle, an object, and an interpretation. All the parts of this triad are needed for the sign to refer. Even so, those three parts hang together in a specific ordinal arrangement, so it is by supposing some of them absent that we arrive at the symbol/index/icon distinction.

For the symbol, if one deletes the interpretation, then the sign as a whole can no longer refer, since interpretation is all that binds the sign-vehicle and the object together (the "deletion" here is simply a prescissive supposition of one thing without another). For the index, if one deletes the interpretation, then the sign-vehicle and the object will remain factually connected, so one needs to delete the object as well in order to extinguish the link that sustains the reference. For the icon, the interpretation and the object can both be deleted and still the sign-vehicle retains its power to refer, because the sign-vehicle's very quality is what ensures its iconic ability.

I want to focus on iconicity, not because it is somehow more important than the other modes of reference, but because its unique character is at the root of many confusions regarding conscious experience. Since an icon stands for its object in virtue of (inherently) resembling that object, it gives rise to a tension. We can

M. Champagne, *Consciousness and the Philosophy of Signs*, Studies in the History of Philosophy of Mind 19, https://doi.org/10.1007/978-3-319-73338-8_5

imagine, at one extreme, a situation where the relata at hand resemble each other so completely that they are in fact one and the same. This is problematic because, in such a case, it would make little sense to speak of a "resemblance."

Twin constraints thus seem to govern iconic representation. On the one hand, the relata at hand cannot be so dissimilar that they are radically alien from each other. Generalizing this, we arrive at Sir William Hamilton's law of homogeneity, which states that "things most unlike must in some respects be like" (Runes 1980, 129). If two (or more) objects are seen to be similar, then the extreme case of such an experience would be that of a complete likeness, merging two things into one. This leads to a concomitant constraint, which is that a given similarity cannot be so pronounced that it effectively collapses the relation between the tokens. Minimally, a split or hiatus must intervene for an iconic bond to be put to use. Accordingly, Hamilton's law of homogeneity must be coupled with the law of heterogeneity, which states that "things the most homogeneous [or] similar [...] must in certain respects be heterogeneous [or] dissimilar" (Runes 1980, 126). Sign-vehicle and object must therefore remain distinct, otherwise they will melt into an undifferentiated unity (indices and symbols are spared this fate, insofar as causal interaction and conventional association both ensure the required rupture).

The following may help to illustrate what is at stake. I am in this room right now. Yet, those surroundings are nevertheless not "in" my mind. The scene I am enjoying from a first-person perspective is therefore, in some sense, a surrogate or stand-in for the genuine article. I want to hold on to this truth. Nevertheless, I am not trapped in some sort of theatre and, so far as I can tell, the experiential display before me is an accurate representation of its object. At the very least, whatever speculative worries might cause me to doubt this would have to be imported from somewhere else. Indeed, if I abide by phenomenological honesty, I have to grant that the experiences I am enjoying are enough "like" their objects to license my confidence that those episodes are a sign of worldly things in a richer and more demanding way than my mere imputation. I do not take the initiative to bind a first-person spectacle with an extra-mental reality. On the contrary, the perceptual signs are motivated in such a way that I do not, by my very act of judgement, bring the likeness into being, but I merely add an explicit grasp to a mind-independent relation that was already there. However, I do not want this to mean that because I have a given thing before me, I own it outright. I don't—and my non-inferential access to it, no matter how thorough and convincing, must never erase that fact. The worldly thing escapes my grasp and gaze by the fact that it is not me. Short-lived episodes of surprise remind me of this (Champagne 2015).

In his *Tractatus de Signis*, the medieval philosopher of signs John Poinsot explicitly considers this tension between sameness and difference in iconicity. Poinsot's Thomistic tradition held that, unless we make our concepts the targets of higher-order reflection, those concepts will efface themselves in regular experience (Dalcourt 1994, 6; Furton 1997, 72, 121). This is because the sign-vehicles spawned by the mind do not work like the sign-vehicles crafted by the hand. "The squeak is physical sound, and the written word is ink on paper. Their signifying function is incidental to their physical being. But a concept does *nothing but* signify" (Wild

1947, 226). Yet, if one agrees with this account, it becomes difficult to justify the contribution of a concept in the first place. Hence, in the course of his inquiry into "what conditions are required for something to be a sign" (Poinsot 2013, 218), Poinsot comes to the conclusion that:

> [T]he more a representation is one with the thing represented, the better and more efficacious is the representation. Yet no matter how perfect, a concept in us does not attain to identity with the represented, because it never attains to this, that it represents itself, but [always rather] another than itself, because it always functions as something vicarious in respect of an object; it always retains a distinction, therefore, between the thing signified and itself signifying. (Poinsot 2013, 228)

Voicing the caveat that such differentiation does not apply to divine revelation (ibid., 233), Poinsot contends that, in all other instances of semiotic mediation, iconic likeness is never complete—and this, as a matter of principle.

Essentially, Poinsot wants to taint similarity so that *a* can never stand for *a*, since the shortcut of simply contemplating *a* would annul the "standing for" relation. Call this "Poinsot's buffer." It requires that "a sign must be more known and more manifest than the significate in the representing, so that in being and in knowable rationale it is dissimilar and [unequal or] subsidiary to that significate" (2013, 217; the square brackets are by the translator). Going over this sentence carefully, the three criteria adopted by Poinsot are that, compared to its object, the sign-vehicle must be (1) more known (*manifestius*), (2) subordinate or less perfect (*inferius*), and (3) dissimilar (*dissimile*).

Although "the criterion that the sign be more known than the signified is clearly what troubles John [Poinsot]" (Furton 1995, 126), I can see why criterion (1) would hold. The "subordination" in (2) seems to be a corollary of (1). However, I think criterion (3) about minimal dissimilarity ultimately undermines Poinsot's realist aspirations. Let me explain why.

Given that human minds can roam across particular locations, Poinsot reasoned that sense impressions, which last only so long, must supply information that survives brief acts of delivery (see Marmo 1987, 117). Indeed, "the sense, by means of the bodily organs, receives a precise impression or image of the sensible qualities, which is impressed directly on the organs" (Pellerey 1989, 87). This is reminiscent of David Papineau's claim that, when a subject undergoes novel experiences, her brain acquires a "stored sensory template" (2003, 359). The moment this happens, the mental content at hand becomes a sign of the thing(s) that spawned it. Because the transmission of contents across the various faculties (Fodor 1983) is thereafter sustained by forces which we would today describe in a neuroscientific idiom, there is a temptation to understand all cognitive processes solely in terms of efficient causation. In the Thomistic account, though, "[t]he sequence of signs from sensible reality to abstract concepts is founded on the validity of the Similitude and on the necessity of the cause-effect relation" affecting our sense organs and brain (Pellerey 1989, 103). Indexicality and iconicity both have a role to play in this account.

Following Aristotle (1993, 42–43), we can use an analogy with a seal of wax to explain how the contents acquired by sensation involve both an iconic and indexical component. The hard seal and malleable substance must at some point collide, so in that respect one can be taken as the index of the other. Still, the wax, once impressed, is an icon of the convex seal and can therefore continue to stand for that object even when detached from its original causal encounter (see Poinsot 2013, 254–255).

Medieval philosophers used this analogy with wax to solve what Peter King (1994) calls "the problem of transduction." A transducer is "a stimulus-bound component" that is "*data*-driven by its environment" (Pylyshyn 1984, 154). An everyday example of a transducer would be a microphone, which turns sound waves into electrical currents (a loudspeaker essentially does the same, in reverse). What is so philosophically interesting about such a device is that, despite the obvious difference between sound waves and electrical currents, something is preserved. Indeed, it is this preservation that prompts us to describe the electricity, not just as a current, but as an electrical signal.

Importantly, induction lets information flow without any symbolic processing. For instance, the taste receptors that cover the tongue are activated by purely chemical means. Chemical receptor sites are more sophisticated than a malleable wax surface or a microphone. Even so, transducers are "computationally primitive in the sense that their internal operation is not considered a rule-governed computation; they are simply performed, or, 'instantiated,' by properties of the biological substrate in a manner not requiring the postulation of internal representations" (Pylyshyn 1984, 154). There is no homunculus in the tongue who somehow "translates" chemical structures into tastes (to evoke Searle 1980). This explains why devices like microphones or loudspeakers break down far less often than computer hard drives.

Despite this lack of translation (or interpretation generally), we have to countenance something that is passed along in the causal encounter, otherwise we would reduce sensation to nothing more than a brute impact between two things. Indeed, if two incorruptible slabs of marble were to strike each other without making any scratch and then resume their previous positions, the short-lived event would not allow for transduction. Aristotle therefore had good reason to pick wax as his example. There might be varying degrees of fidelity in the information that is passed along. But, if the signet-ring has, say, an S-shape on it, then transduction ensures that the wax will take on, if not an identical S shape, then at least an S-like shape.

One consequence of this iconic fidelity is that, when a transducer is functioning properly, one can reliably infer its inputs merely by knowing its outputs. However, in keeping with the idea that icons and indices survive the deletion of interpretants, inferences are not what hold together transduction. For instance, one can tell, merely by considering a microphone's electrical signal, whether a person was singing softly or loudly. But, crucially, one's ability to make this inference is not what converts specific rates of airwaves into matching rates of electrons. The inferences thus piggyback on a non-inferential process.

There is no denying that, like the seal pressed on the wax (or the wax pressed on the seal?), our experience of the world involves impacts of sorts. To that extent, sensation is indexical. Poinsot has no problem with this. However, iconicity gives him pause, because an icon that would be too pronounced would fail to be a sign. Yet, if the stand-ins that result from situated impacts can also link minds to the world in virtue of their very quality, why would anyone committed to realism refuse this?

I think that what prompts Poinsot to add his *dissimile* clause is Aquinas' thesis that "[a]lthough it is necessary for the truth of cognition that the cognition answer to the thing known, still it is not necessary that the mode of the thing known be the same as the mode of its cognition" (*Summa Contra Gentiles*; translated by Rasmussen 1994, 417). A default recommendation to tease these two elements apart can certainly benefit inquiry in most instances. The problem, however, is that in iconicity *what* we know is precisely *how* we came to know it.

Peircean semiotics allows us to make sense of this. Peirce did not know of Poinsot's writings, but he knew the medieval philosophical literature quite well (Boler 2004; Beuchot and Deely 1995; Tiercelin 2006). Indeed, "[w]hatever agreement we find between Peirce and Poinsot is based upon them independently discovering the same things about the nature of thought-signs" (Maroosis 2003, 157–158). Poinsot and Peirce thus partake (from afar) in a common semiotic project. "If C. S. Peirce can be said to give us a chemistry of [...] sign compounds, John Poinsot, suitably revised, gives us the basic physical laws of motion that bring sign, object, and mind into relation" (Murphy 1994, 569).

Peirce agrees with Poinsot that, since a sign stands for something else, "one and the same thing never represents itself; for this identity cancels the rationale of a sign" (Poinsot 2013, 234). In fact, in manuscripts that have only recently been published, Peirce contemplates the same problem addressed by Poinsot—and comes near to giving the same answer: "If it were in all respects a perfect icon it would be indistinguishable from and for all intents and purposes the very same thing as its Object. It thus seems to be of the very essence of an icon that it should not be perfect [...]" (Peirce 2015a, 658). Interestingly though, Peirce differs from Poinsot by making room for a complete merger in what he called "Firstness," which is the state one gets when one subtracts a relation between relata so as to obtain only a lone relatum (the term "relatum" thus becoming a misnomer). As a logician, Peirce (1998, 186–194) accepts that the triadic relations involved in semiosis must subsume simpler (dyadic and monadic) ones—even if this means that, below a certain point, there cannot properly be a sign (Santaella 2003, 49–50).

The disagreement between these major historical figures, then, turns on whether semiotic theory should countenance such a state of unbroken unity, or whether it should make division-into-two the most basic situation attainable. Poinsot thinks that, in analyzing the sign, we should stop at two things, before a merger between sign-vehicle and object is reached. Peirce thinks that we should push our analysis all the way down to one, where there is a merger. I favour Peirce's stance.

Although Peirce develops a distinct phenomenology, the notion of the "phenomenal" suggests a "phenomenalism" which he, as a scientific realist, found repugnant (see Ransdell 1978). Qualia, in his system, do not occlude the world. On the contrary, because they have an inalienable iconic power, qualia can reveal the world. I thus believe that, properly handled, Peirce's account can demystify Aristotle's claim that, when things go well, "knowledge is identical with its object" (1993, 60).

There is more at stake here than individual knowledge. In Chap. 4, I argued that two people in a vast white expanse can use prescission properly to experience the same colour tone. If, however, our semiotic constitution is such that we can approximate but never fully match a given quality, then we can never use this quality to fully share our state(s) of mind (see O'Dea 2002, 177). We may use indices to ensure that our behaviours and utterances co-vary, but trying to achieve genuine empathy and agreement by jointly attending to evidence becomes an inherently flawed aspiration. This would be a major concession.

Now, a standard approach to truth sees it as consisting of truth-bearers on one side, truth-makers on the other side, and a truth-relation between the two. Whatever "truth" we get in iconic reference reminds us not to take this model for granted, since in our most proximate experiential dealings with objects, truth-bearer and truth-maker are one. Indeed, correspondence theories of truth can involve either "correlation" or "congruence" (see Kirkham 1997, 119–120). Congruence uses similarity, but correlation does not require that the veridical correspondence involve any kind of isomorphism between truth-bearer and truth-maker. Poinsot espouses correspondence as correlation, whereas Peirce thinks that correlation can subsume congruence.

Peirce's argument could be put like this. If any two objects X and Y are similar in some respect, then X should possess that "respect" all on its own. Hence, were Y to vanish, X would retain the feature that made a comparison by similarity possible. Obviously, this applies to Y too. Yet, when we focus only on the relevant quality, we make ourselves unable to ascertain whether it is X or Y that is the quality's bearer. Hence, at the proper level of analysis, whatever makes Y and X similar to each other is indifferent to where it is found. This means that one should not worry about (much less accuse theorists of) conflating physical presence and cognitive presence (Levine 2007, 163) since, at the proper level of analysis, there is simply nothing to "conflate."

This may be what Peirce's friend William James had in mind when he noted that, approached from a certain perspective, "the sensation as 'sign' and the sensation as 'object' coalesce into one, and there is no contrast between them" (2007, 243). Like the empty Sheet of Assertion of Peirce's diagrammatic logic (Peirce 2015b, 900), James invites us to consider a white sheet of paper and note that, looking only at that unbounded expanse of colour, "[t]he thought-stuff and the thing-stuff are here indistinguishably the same in nature […]" (1975, 31). In the current philosophy of mind literature, Matjaž Potrč defends something similar when he argues that conscious experience presupposes an "original intertwinedness" of subject and object (2008, 110–111). I believe a tenable account of consciousness should countenance such an iconic bond.

The basic idea motivating Poinsot's buffer remains correct: lack of difference does indeed rob the minimal relation (and sense of direction) that allows us to properly call something a sign. But, as a logician specialized in cataloguing relations, Peirce recognized that the complex triadic relations involved in semiosis subsume simpler (dyadic and monadic) ones. Hence, there should be no logical obstacle to supposing some parts of the sign absent.

Most of the confusion regarding phenomenal qualia in philosophy of mind stems from the fact that such abstract deletions can be carried past the point of numerical distinctness. In the most impoverished state, signification (and thus cognition) vanishes. A quality would stand for anything like it, but that potential lies dormant if all we have before us is the quality itself. In that regard, icons are transparent and opaque, depending on how we view the situation. This vacillating semiotic/non-semiotic status is rendered soluble by keeping track of how many things one is countenancing in a given analysis. The Peircean category of Firstness is meant to remind us that, with one sign-vehicle in complete isolation, there can be no sign-action.

When, for whatever reason, the stream of consciousness which associates one thought to another ceases and is submerged in a unique thought, the referential potency inherent in that paused qualitative state cannot be actualized. To actually refer, something *else* must enter the picture. Peirce writes that "in contemplating a painting, there is a moment when we lose the consciousness that it is not the thing, the distinction of the real and the copy disappears [...]. At that moment we are contemplating an *icon*" (1931–58, 3.362). This is the state I am in when I transparently take my experience to be what is experienced, when I partake of my environment so completely that I lose it and myself.

Of course, aside from pharmaceutically-induced vegetative states, such a qualitative merger of "I" and "it" is rarely consummated; otherwise, as Poinsot put it, this "cancels the rationale" of the sign (2013, 234). Still, Peirce is distinguished from Poinsot in making allowances for this possibility. Indeed, Peirce concludes that, in theory, pure "[i]cons are so completely substituted for their objects as hardly to be distinguished from them" (1931–58, 3.362). In short, careful study of the conditions for the possibility of signification reveals a ground-level where similarity becomes so pronounced that it eradicates that with which the quality is similar. Putting these results in the technical terms used earlier, we could say that while actual sign-action is indeed constrained by Hamilton's twin laws of homogeneity and heterogeneity, such actual semiosis logically implies a possible semiosis free from the constraint of heterogeneity.

Poinsot holds that "a sign must be dissimilar [to its significate], [...] otherwise, if it is equally manifest, there is no reason for this to be a sign of one thing rather than a sign of some other thing [...]" (2013, 218; the square brackets are by the translator). If, however, the ground of any experience is likeness to the point of complete merger, then when one subtracts the divide between sign-vehicle and object, one obtains a communion between knower and known that is as close as close can be (Ransdell 1979).

Do Poinsot and Peirce really differ on this point? John Deely suggests that the Latin term of *idolum* that we find in Poinsot is analogous to the contemporary Peircean notion of icon (Deely 1982, 115; see also his footnote in Poinsot 2013, 241). However, since Poinsot's system does not contain anything analogous to Peirce's category of Firstness, I am not entirely convinced that their respective texts support the agreement that Deely and others (e.g., Maroosis 2003, 160) look for. I think the closest Poinsot comes to a pure icon in his *Tractatus de Signis* is when he mentions (only once) a *similitudo virtualis* (Poinsot 2013, 258), which is presumably the modally prior "principle *whence arises* a formal similitude and formal awareness" (ibid., 258; emphasis in original). However, the term "virtual" will not be helpful until it is unpacked into something more tangible. The Peircean analysis does just that: to be an icon is to enjoy a monadic relation with an object, which is to say no relation at all.

Poinsot, as I mentioned, exempted a privileged subset of cases from his buffer. Specifically, he held that "the Divine Word is excluded from the rationale of sign" (2013, 233) because it supposedly "attains to identity with the represented divine essence" (ibid., 228). Lifting the buffer was intended to spare scriptures and sacraments from being mere symbols. Interestingly, Joseph Ransdell (2003, 229–231) speculated that the present discomforts regarding iconicity might be a hold-over of religious controversies. Are there any religious implications in the idea of iconicity? Mark Johnston (1992) proposed the term "revelation" to capture our direct appreciation of qualities like colours. That choice of word certainly courts non-secular readings. However, all that Johnston wants to convey is that "[t]he intrinsic nature of canary yellow is fully revealed by a standard visual experience as [*sic*] of a canary yellow thing" (1992, 223). Surely this is not far-fetched.

Of course, on the prescissive analysis I have championed, focusing on a yellow quality at the expense of all relations is by no means a "standard" activity. To get there, one must artificially trim away quite a bit of lived experience. Still, given that the qualitative unity we find in iconic signs logically underpins the human cognitive situation as a whole, I have endeavoured to account for it in a way that does not veer into philosophical (much less theological) extravagance.

I should underscore just how close Poinsot's semiotic hugs the asymptotic limit of complete likeness. He acknowledges that "in intentional or representative existence the formal sign is said to make one thing with the object, not only as do those things which coincide in one common rationale, but rather because it totally contains and represents the numerically same being that is in another" (2013, 233–234). But, he immediately adds: "this fact itself supposes that the representing and the represented are distinct [...]" (ibid., 234). If Poinsot's buffer is in place, then strictly speaking his secular philosophy of signs does not contain any icons, only symbols and indices. Very similar indices perhaps, but indices nonetheless.

I think that, by preserving a minimal qualitative dissimilarity between mind and world, Poinsot deprives the semiotician's quiver of a useful arrow. One of the tenets of Poinsot's account is that, insofar as it acts as a sign, a concept "as such does not stand in need of some scheme, linguistic or otherwise, to relate it to reality" (Rasmussen 1994, 410). I agree. However, I side with Peirce in holding that this is because, on some level, a sign-vehicle *just is* its object. A lot of things fall into place once we realize this. Indeed,

> We may understand Quine's criterion of ontological commitment in Peircean semiotic terms as an attempt to place the full burden of representing reality onto *indexical signs*. This leads philosophers with realist sympathies to feel they need to ask a raft of questions of the form: "Does term X [e.g. ethical or aesthetic predicates, number-terms...] denote a real object?" If we recall that indexical signs pick out sign-independent particulars, it often seems hard to answer "yes" to this question for key terms in manifestly important human discourses [...]. (Legg 2013, 16)

Qualia are not captured by indexical signs, so they are a casualty of this assumption. I think, though, that "any project of explanation that applies to consciousness the empirical methodology of the experimental sciences risks falsifying or omitting entirely the interpretive kind of access that we have to our own consciousness [...]" (Livingston 2004, 229). Enlarging the menu of referential options to include icons is one way of avoiding this omission.

Icons are ideal informational transducers because they involve no transduction; "[a]nd this means that philosophers do not have the task of explaining how we get *from* our experience *to* its external object" (Hookway 2007, 68). This lets us reject the view, expressed by Bertrand Russell, that "[t]he starry heaven that we know in visual sensation is inside us" whereas "[t]he external starry heaven that we believe in is inferred" (2015, 57). We can see things as they are.

Interestingly, a similar conclusion is currently being reached in analytic discussions. In a statement that comes very close to iconicity, Papineau observes that "in phenomenal thought the conscious referent seems to be present in the thinking itself, without any veil between subject and object" (2006, 104–105). One notable consequence of this is that "[w]hat it's like to focus phenomenally on your visual experience of the bird is no different from what it's like to see the bird" (Papineau 2007, 124). I could not agree more. However, the Russellian in Papineau soon wakes up, so he discards these results out of some prior conviction that "[p]henomenal thoughts do not have any magical power to reach out and grasp their objects transparently" (Papineau 2006, 105).

I do not detect any magic here. Papineau's incredulity arises only on the assumption that "[p]henomenal thoughts, just like any other intentional states, gain their referential powers from causal and historical relations" (2006, 105). If there is any moral that I hope to impart in this book, it is that not all modes of reference conform to this causal model.

5.2 Un-Lockeing a Coloured World

The pure icon discussed in the previous section is a theoretical ideal—the asymptote of a likeness bereft of any alterity (qualitative and numerical). It nevertheless gives us a way of picturing knowledge at its best. Error is possible. But, on the view I advocate, "there is no ontological gap between the sort of thing one can mean, or generally the sort of thing one can think, and the sort of thing that can be the case. When one thinks truly, what one thinks *is* what is the case" (McDowell 1996, 27; see Dodd 1995). Schooling in philosophy can make this direct realism look sanguine, but in truth it brings thinking back to where we started. Indeed, "[a]sk a child where the green of the lawn is and he will point out *the lawn* [...] not to the eye or head of the viewer" (Stroud-Drinkwater 1994, 347; see Kelley 1988; Allen 2016).

Peirce's iconic merger can vindicate this common sense view. However, "the most popular opinion, at any rate among color scientists, may well be the view that *nothing* is colored—at least not physical objects in the perceiver's environment, like tomatoes" (Byrne and Hilbert 2003, 3). It seems that, somewhere along the way, inquiry has strayed from what Wilfrid Sellars (1991, 1–40) called the "manifest image."

This distrust of qualitative experience owes much to John Locke. Locke invented the word semiotics (Deely 2003), but he also invented the inverted spectrum (Locke 1825, 279–280) and, according to some (e.g., Balibar 2013), he invented consciousness too. Now, John Deely has repeatedly warned that Locke is an ambivalent historical figure, since there is an "antinomy between the actual point of view adopted at the beginning for the *Essay* as a whole and the possible point of view proposed at its conclusion" (1990, 114; see also Deely 1994). Let me therefore focus on one salient tension between empiricism and philosophy of signs, namely Locke's demotion of qualities like colours, smells, and tastes to a "secondary" status.

Locke claimed that solidity, extension, figure, motion, rest, and number "really exist in the [physical] bodies themselves;" but he held that feelings like sweet, blue, and warm "have no resemblance of them at all" (1825, 76). This division follows from the view that anything real must, at bottom, be composed of "atomicules [that] all alike act mechanically upon one another according to one fixed law of force" (Peirce 1998, 186). Such causal events are held to have a basis in fact, "[b]ut as for Qualities, they are supposed to be in consciousness merely, with nothing in the real thing to correspond to them [...]" (ibid., 187). Locke's account of consciousness requires a fair bit of reconstruction (Weinberg 2016). Still, if Locke is right that many of the qualities we experience are suspect, then that does not bode well for iconicity, since that mode of reference employs colours and smells just as easily as it employs shapes. Since I do not regard any experiential quality as suspect, I want to reinstate the experiences expelled by Locke's account. To do that, I will use historical criticisms and philosophical arguments. Let me begin with the historical criticisms.

Locke cordoned off experiential qualities because, no matter how obvious they may seem to us, they were not mentioned by the physicists of his time. Like W. V. O. Quine (1966, 151), who proclaimed that "philosophy of science is philosophy enough," Locke held that "it is ambition enough to be employed as an under-labourer in clearing the ground a little, and removing some of the rubbish that lies in the way to knowledge" (1825, x–xi). Yet, Locke never actually argued for this deferral. That the scientific theories of his time were thriving was presumably enough to motivate Locke's meta-philosophical stance. Surely, if one is going to assign philosophy a subordinate role, the demotion requires some justification.

Be that as it may, the Lockean vision of "a world made out of ultimate little things and collisions amongst them" (Ladyman et al. 2007, 23) continues to figure prominently in philosophy of mind. Frank Jackson, for example, claims that all our knowledge of the mind "comes back to causal impacts of the right kinds" (2004, 418). Yet, if everything the mind knows comes by way of collisions between two things (i.e., Secondness), then it follows that qualities considered apart from all other things (i.e., Firstness) will appear suspect. I do not want to let go of the idea that perception involves at least two things (usually a perceiving subject and a perceived object). Still, I am trying to call attention to another possibility, where seeing something is being one with that thing. If the vocabulary of "being one with" seems strange or high flown, then that just goes to show how much revisionist work is needed.

Peirce was aware that many follow Locke in taking physics as their metaphysics. As he notes, "[Henri] Poincaré would have us write down the equations of hydrodynamics and stop there" (Peirce 1998, 187). Like Locke, Peirce admired natural science. Unlike Locke though, Peirce believed that natural science needs to countenance formal causes alongside efficient causes. For Locke, it is "manifestly by impulse" that ideas are produced in us (1825, 75). For Peirce, this is not at all obvious. In what has been described as his most "scholastic" passage (Deely 2009, 268), Peirce wrote that "[t]hat which is communicated from Object through the Sign to the Interpretant is a Form" (Peirce 1998, 544n22). Causality may be central to perception, but it is not the only way that minds can be tethered to the world.

Medieval philosophers distinguished three senses of sameness, "namely, identity (sameness in substance), equality (sameness in quantity), and similarity (sameness in quality)" (King 1994, 131n12). Locke knew of this scholastic account (see Jacovides 1999, 463), but he clearly distrusted sameness in quality. So, whereas I stress the centrality of iconicity in uniting mind and world, Locke introduced his distinction between primary and secondary qualities "so we may not think (as perhaps usually is done) that [our ideas] are exactly the images and resemblances of something inherent in the subject; most of those of sensation being in the mind no more the likeness of something existing without us [...]" (1825, 74). However, Locke never tells us why we need to be saved from the supposedly quite wrong belief in mind-world likeness. Since, by his own admission, Locke cannot get behind the appearances to check how things really are, one is left wondering what licenses his confidence about a lack of similarity. One cannot infer from the premise that "most" ideas do not resemble external things to a conclusion that none resemble them.

Locke did not make this fallacious inference outright, since he granted that some of our ideas get to enjoy such a resemblance. Indeed, "[w]hatever Locke means by 'resemblance,' the relation is one of the two ways that he believes that ideas can represent external objects. The other way is by brute causal connection" (Jacovides 2007, 107). This looks like the enlarged semiotic repertoire that I defend in this book. The only problem, though, is that Locke lets only "primary" qualities enjoy a resemblance to their objects. So, if one takes the idea yellow triangle, one can trust that the triangularity resembles something out there, but the yellowness is met with distrust.

Locke likely got this distrust of "secondary" qualities from Galileo (see Aho 2014, 94–97). However, Galileo's distinction is arguably a reinvention of the scholastic distinction between "common" and "proper" sensibles. We rarely use that medieval jargon today, but it is worth revisiting. Proper sensibles (like the taste of chocolate) can be detected by one sense alone, whereas common sensibles (like the rectangular shape of a chocolate bar) can be detected by more than one sense (in the case of rectangles, by sight and touch). It is important to underscore, however, that this scholastic distinction does not entail any skepticism. Proper sensibles can be verified, they just cannot be cross-verified by other senses. Hence, scholastic thinkers used the possibility of redundancy to sort experiences, but there was never any question that we experience the full range of experiences, tastes and colours included.

As the notion of common sensibles shows, different senses can sometimes overlap in what they tell us. However, it is unclear to me why a message coming from one source should be deemed less reliable or trustworthy than a message coming from two (or more) sources. If a cunning devil can fool each sense, then surely it can fool two (or more) senses working in tandem. Despite this, Locke followed Galileo in adopting the primary/secondary distinction in its sceptical variant. Locke certainly wasn't a fan of the religious establishment that placed Galileo under house arrest. Compared to the predictive power achieved by the new generation of "natural philosophers," the proper sensibles countenanced by scholastic philosophy seemed to accomplish nothing. He may have been right. Even so, I do not think that Locke's philosophical response was appropriate.

As a physicist and astronomer trying to predict the motion of bodies, Galileo did not have much use for qualities that cannot be quantified. Despite his admiration of Galileo, Locke was never quite comfortable with the idea that everything is ultimately mathematical in nature. As an empiricist Locke held that exposure to things is needed in a way that no amount of theoretical insight can supersede. In a passage that anticipates Jackson's (1982; 1986) knowledge argument, Locke writes that "if a child were kept in a place where he never saw any other but black and white, till he were a man, he would have no more ideas of scarlet or green, than he that from his childhood never tasted an oyster, or a pine-apple, has of those particular relishes" (Locke 1825, 52–53). As healthy adults, we are unlike Locke's secluded child or Jackson's secluded neuroscientist in that we have been flooded with many different (and many similar) experiential qualities. It is too late to put ourselves in a state of qualitative deprivation; were we to succeed in doctoring the right kind of isolation, the sudden drop or lack would be noticed—which is precisely what Locke's child and Jackson's neuroscientist are not supposed to notice. As philosophers, though, it seems we can disown all that the world has taught us and call into question the very existence of our conscious experience. Or at least vast swaths of it.

Locke's colourless worldview fit nicely with the Newtonian physics that he admired, but it burdened him with explaining why/how secondary qualities emerge from (or are epiphenomenally attached to) the primary ones. Locke gave a story about tiny "corpuscles" hitting the sense organs, but that story never successfully got past the fact that "matter, incogitative matter and motion, whatever changes it might produce of figure and bulk, could never produce thought" (1825, 477). Working out the consequences of his starting assumptions, Locke was forced to conclude that conscious experience of qualities like colours may accompany material events "merely by the good pleasure and bounty of the Creator" (Locke 1825, 403–404).

Locke's deistic scaffold eventually withered, but his sense of mystery did not. By the nineteenth century, we find T. H. Huxley writing that conscious experience of qualities like colour is "as unaccountable as the appearance of the Djin when

Aladdin rubbed his lamp in the story [...]" (1866, 193). In the late 1980s, Colin McGinn turned this mystery into a bona fide stance of philosophy—"mysterianism"—when he revamped Huxley's phrase and wondered "[h]ow can technicolour phenomenology arise from soggy grey matter?" (1989, 349).

What is the likeliest explanation for this lack of explanation? That the world contains some weird discontinuity which makes our own minds seem out of place—or that we have taken on a fundamentally mistaken worldview? Nagel is probably right that "almost everyone in our secular culture has been browbeaten into regarding the reductive research program as sacrosanct, on the ground that anything else would not be science" (2012, 7). Indeed, many philosophers see themselves as "defending the hegemony of modern matter against the mysteries of mental substance and of mind/matter interaction" (Crane and Mellor 1990, 186). Yet, this campaign to rid philosophy of mind of mysteries ends up generating even greater mysteries.

Some hold that qualitative experiences emerge at a certain level of complexity (Carruthers 2000, 237–238; Deacon 2011, 530–531; Rosenthal 2010). I do not find such accounts plausible. The edifice being erected has no qualitative ingredient in it, but once we reach an upper floor, qualities appear—somehow. While I am hesitant to endorse the panpsychism recently resurrected by David Chalmers (2015) and Galen Strawson (2006), I agree with them that, if we want to explain the qualitative dimension of consciousness in a non-miraculous way, we must weave experiential properties into the fabric of reality. I therefore endorse the following statement:

> Panpsychism rests upon a fairly strong version of the principle of sufficient reason, which it applies across the board, metaphysically as well as epistemologically. Unlike emergentism, panpsychism doesn't tolerate metaphysical discontinuities—it insists that high-level entities issue from bottom-level entities. Unlike physicalism, panpsychism doesn't tolerate epistemological discontinuities—it insists that high-level entities issue *intelligibly* from bottom-level entities. (Lewtas 2013, 40)

Interestingly, if we go to the original source, we find that the famous quote by Huxley actually takes consciousness to be "as unaccountable as the appearance of the Djin when Aladdin rubbed his lamp in the story, *or as any other ultimate fact of nature*" (1866, 193; emphasis added). Like the passage where Peirce (1931–58, vol. 4.537) explicitly says that a tone "can neither be called a Type nor a Token," the final portion of Huxley's passage is often left out (see for example Bayne 2004, 361). This omission once again undercuts possible solutions, since Huxley's full claim hints at two responses: magical emergence—or simply countenancing consciousness alongside the other primitive facts of nature. I prefer the latter response.

I am not alone. As Pär Sundström reports, "[a] number of philosophers have argued that Galileo was wrong [...] in taking the *qualities* we experience in physical things to be properties of the mind" (2007, 128). Backers of this Galilean reversal include (but are not limited to) Byrne and Hilbert (2003), Harman (1990), Hilbert (1987), and Tye (1995, 2000). Everyone seems to be gravitating toward a shared conclusion by independent reasons/reasoning. Still, the emerging consensus seems to be that "the only plausible way that a Materialist can deal with the secondary qualities is completely to reverse the whole programme started by Galileo, a programme that has persisted for so long" (Armstrong 1999, 124).

Locke thought that if persons "should really have different ideas, I do not see how they could discourse or argue one with another" (1825, 109). I agree with John Deely that the moment Peirce and others took this seriously and began "to think that their experience of communication [...] as a proper starting point of philosophy, the remaining days of classical modern thought were numbered" (2001, 539). Semiotics—the project hinted at by Locke—turns out to be the best paradigm for undermining Locke.

5.3 The Argument from Subjective Variation

When subjects discuss a "secondary" quality like the taste of cilantro, they are engaged in an act of interpretation. Hence, a lot of what Donald Davidson has written about "triangulation" (2001, 212–220) is applicable. The three tips of the triangle are two persons and one object under discussion (in this case, a sprig of cilantro). Using language, interlocutors are trying to agree about what they intend. Philosophy of signs and philosophy of language are natural allies here. But, if my earlier chapters are right, then such an exercise in triangulation will never succeed in fully capturing qualia unless the passage from conventional signs (symbols) to causal signs (indices) culminates with qualitative signs (icons).

In order for interlocutors to whittle discourse down to an iconic point, a lot of situated pointing and charitable chatting may have to occur. Beyond a certain point, though, nothing more can be said (Wittgenstein 2002, 189). As I see it, misgivings about "secondary" qualities rest on the assumption that, when the exercise becomes silent, the exercise has failed. I see no reason to draw that conclusion. In my account, "[t]he verbal argument is at most only stage setting; the heart of the drama is the invocation of experience and, indeed, the attempt to register accurately the felt force of relevant experience" (Colapietro 2010, 11).

I do not want to downplay the complexity of a communicative exercise or the rarity of a full agreement. Many of the difficulties that accompany triangulation stem from the fact that, when people discuss the taste of cilantro, they are not just making claims about cilantro, but also about cilantro-insofar-as-it-interacts-with-sense-organs-here-now. The taste reported is therefore, implicitly, a report about the functioning of our bodies and our position in the world (O'Dea 2002). A full description of what one claims when one says that "This cilantro sprig is bitter" would thus have to include facts about the temperature of one's mouth, allergies one may have, one's blood sugar levels, the number and distribution of receptor cells on one's tongue, psychological priming effects one may be biased by (e.g., being told right before "Wait until you taste how bitter this is..."), or whatever else might affect how one experiences things. No one—least of all Locke—has experimentally controlled for all these variables. I thus find it hasty to conclude that, because there can be disagreements between persons, tastes are subjective.

Interestingly, when it comes to other features, we do not jump to this subjectivist conclusion. Consider the fact that a small child cannot lift me but a strong adult can. Clearly, it would be erroneous to drop the relations involved and hold that I somehow have two clashing predicates, "liftable" and "unliftable," and that since I cannot be the site of both predicates, debates about my liftability are a matter of personal preference. Likewise, a sprig of cilantro might taste very differently to two people, but that is no reason to squeeze two mutually-exclusive qualities into one poor herb so as to infer that its distinctive flavour cannot possibly be "out there."

If I am right, then triangulation, pursued with enough care, can eventually result in genuine agreement. This is, after all, the optimistic asymptote that we chase when we try to convey "primary" qualities. An engineer trying to communicate a complex shape to another person does not stop at the first sign of a misalignment between speaker and hearer. Likewise,

> We do not universally agree in our judgments about what is red or salty. But when there is disagreement, we do not blithely continue to maintain our own views without hesitation. The fact that others report seeing red where you saw green, or tasting saltiness where you tasted none, makes you less confident in your own color or flavor judgments. It makes you suspect that the lighting is funny, or that you are ill or under the influence of a drug, or that your perceptual equipment is defective (as it is in color-blind people). To insist without further investigation that your own judgment is right, and that the other's is wrong, would be rash and unwarranted. (MacFarlane 2014, 5)

It is normal that scientific practitioners seeking consensus should limit themselves to observations that they can talk about with consistent inter-subjective agreement. But, a disciplinary predilection for communicability does not mean that incommunicable experiences merit ontological or epistemological demotion.

In my account, minds can eventually converge/merge by employing symbols, indices, and icons. The unique semiotic service rendered by each kind of sign cannot be replaced. It may be that "[c]urrent perceptual science, even when it deals with qualitative aspects of experience, almost exclusively explains them in terms of the stimulus *quantitatively* understood as a collection of objective measurements" (Albertazzi 2012, 9). However, the scholastic one-channel/multi-channel distinction neither entails nor mandates privileging common sensibles—any more than "This cloth can be washed by only one machine" entails or mandates "This cloth cannot be washed."

Of course, eliminative materialists (e.g., Churchland 1996, 207) argue that, if one truly understands the relevant physics and neuroscience, one will see that a quality like red is in fact a wave length. Alas, my experience of colour persists—despite all the science courses I have taken. Faced with this indispensability, one can hold that "our perceptual experience is incurably infected with illusion" (Shoemaker 1994, 296)—or one can try to tell a story that rescues the experiences from being systematically confabulated. In this book, I work on the second option.

5.4 Colours and the Soggy Grey Brain

I have been trying to remedy scruples about colour qualia using historical scholarship and philosophical arguments. Could science help to vindicate our experience of qualities like colour? A common reflex nowadays is to look to the brain for (all the) answers. The crudest form of investigation would therefore be to crack open a skull to see what is in there. Interestingly, researchers have done just that.

The brains of macaque monkeys were injected with a chemical compound that has the property of reacting with activated brain regions. Like the famous brainwashing scene from the movie *A Clockwork Orange*, the monkeys' gaze was fixed so that they had to stare directly at a grid-like image. After about 30 minutes, the deoxyglucose stains set, allowing researchers to detect which portions of the brain were activated by the visual stimulus. A tissue section of the macaque striate cortex was surgically removed and flat-mounted for examination. The results obtained were astounding, revealing a sort of imprinting of the grid (see the figure reproduced in Tootell et al. 1982, 902). "Like pictures, adjacent neural populations in these areas corresponds [*sic*] to adjacent boundaries or surface points in the objects they represent" (Prinz 2002, 31). Nelson Goodman (1976) may have been led by his love of abstract art to philosophically repudiate resemblance, but the results obtained by Tootell et al. (1982) clearly show a portion of the world tattooing its likeness on a portion of the brain.

Considered as an index, it matters that the grid-like pattern travelled (via transduction) from the external visual stimulus to the cortex. Given the direction of cause and effect, we naturally take the brain imprint to be a sign of its stimulus. If, however, we consider the grid patterns as icons of each other, there is no order of priority. Indeed, these two patterns will resemble each other, even when no further thought (by us or the monkey) represents them as doing so. This shows that, "[e]ven before having a fully functional semiotic consciousness, our body is not mere presemiotic matter, but a highly complex semiotic system, endowed with […] the capacity to make subtle distinctions and respond in competent and meaningful ways to salient environmental stimuli" (Violi 2007, 84). Localized calluses on hands or feet could have shown us this already, but since we are more likely to credit the brain with knowing, discovering a trace of the world on that soggy grey organ is more reassuring.

In order for a brain-owning organism to cope with its environment, the growth of signs which ensues from its primitive bodily engagement must eventually switch to some sort of symbolic encoding. The claim of Peircean semiotics, though, is that the chain of signs does not begin with symbols. Thomas Sebeok and Marcel Danesi (2000, 11) call this the dimensionality principle: iconicity precedes indexicality, and indexicality precedes symbolicity. Minds tied to the world by the dimensionality principle do not face any problem grounding their symbols (Harnad 2002).

In keeping with the dimensionality principle, the sequence of cognition does seem to begin with icons. Trivially, the outer-most layer of the eye (where we put contact lenses) reflects whatever it is exposed to. Anyone who has cheated at poker by seeing cards on a player's eye balls has been acquainted with this informational coupling. These icons can be useful to outsiders, but they are of little use to the subject who owns the eyes. So, moving further inside the body, anatomy teaches us that the inner wall of the eye has a light-sensitive membrane. Here again, thanks to photoreceptors, iconic patterns are being preserved. If, for instance, the top half of a shown image is darker than the bottom half, this relation will be maintained on the membrane, albeit in a reversed fashion. Because this permutation is simple, we have access to the "code" that allows us to go from one body of data to the other. If you show a biologist a stimulated retina, she will have little difficulty reverse engineering what the stimulus was. However, as we move deeper, along the optic nerve to the visual cortex and beyond, more elaborate encoding occurs. By the time we get to the brain, the link tying neurons to the world is overwhelmingly complex, like trying to read an unknown language.

The results obtained by Tootell et al. (1982) nevertheless prove that an iconic relation is being preserved past the retina. The discovery of imprinted patterns on the striate cortex thus reduces the epistemic risk of our inferences about consciousness. Now, these results are not the norm, since most of the information stored in a brain bears little or no resemblance to the objects and events which shaped it. Still, rationality demands that we give weight to similarity relations. If, say, I show you a picture of man with a red stain on his shirt, you are more entitled to infer that he spilled his glass if the colour of his wine is red than if it is white. Likewise, the match in quality that we find in a monkey's brain adds some security to our inferences about that monkey's conscious experience. Judged by the standard of deduction, these inferences remain inconclusive. But, all other things being equal, co-variations with a matching quality are more secure than co-variations alone.

Assuming that the experiment in question could be replicated in humans with the same results, switching back and forth between similar patterns should diminish our doubts that the subject did in fact experience the image it was shown. However, this added sense of security only goes half the distance. Indeed, the grid-like chemical stains count only as a "primary quality" in Locke's sense. Somewhere along the way, "secondary qualities" like colours have been lost.

The researchers I am discussing did not bother to track if and to what extent the colour(s) of their stimulus might have also been conveyed alongside shapes. Discovering that there is "a systematic relationship between ocular dominance strips and cortical magnification" (Tootell et al. 1982, 902) is worth writing about, but apart from incidental colouring by chemical stains, we know in advance of any experiment that the brain is and will remain a grey and soggy thing. Yet, as a philosopher I want to ask: would the "technicolour phenomenology" (McGinn 1989, 349) of our experiential lives seem less mysterious if we could find colours in the brain?

Let me toy with that possibility and see what it can teach us. Imagine that a gland, dubbed "the chameleonic gland" by its discoverers, was found somewhere in the brain, near (and suitably connected to) regions known to be responsible for vision. Under surgical conditions, we could observe this gland acting like a chameleon by becoming, say, red, whenever the subject was shown a red stimulus (and so on for the rest of the visible colour spectrum). Unlike Descartes' pineal gland, there would be no disconnect: the stimulus would be red, the brain would be red, and the subject would report seeing red. Since correlation is the abductive gateway to causal inference, the robust match in colours would increase our confidence that the quality at hand is shared all the way to first-person experience.

As we have seen, going back and forth between similar grid-like shapes increases our confidence that a subject consciously saw the shape in question. From a rational standpoint, then, the same increase in confidence would apply if we could go back and forth between two similar colours. Yet, when it comes to shapes, I know of no one who is willing to say that, had researchers not gathered evidence of a grid-like shape in the striate cortex, the ontological standing of shapes would have been suspect. Even before the scientific discovery of shapes in the striate cortex, our ordinary experience of shapes sufficed to justify our belief that shapes exist in the world. By parity, I argue that we need not wait for a discovery of colours in the brain to believe that colours exist in the world.

For many, this suggestion engenders a recoil. Taking "primary" qualities like shapes at face-value is respectable, but treating "secondary" qualities like colours the same way is "naive" realism. I am trying to deconstruct the double-standard that prompts this recoil. A healthier reflex is to ask: what would convince an eliminative materialist that colour qualia exist? The chameleonic gland is my closest answer. Yet, the more we contemplate this "smoking gun" proof of qualia, the more we see that such a revolutionary finding would be somewhat banal. After all, staring down at a red gland in an open skull is not that different from staring down at a red strawberry in an open basket. My claim, then, is that unless we consider the boundaries of the skull to be paramount, evidence for the existence of colour qualia is available right now.

A growing number of philosophers agree. Joining their rank, I have considered two cases—one real, the other fictional—of the brain resembling its environment. From a rational standpoint, inferences about conscious experiences are more secure when they proceed from an iconic and indexical bond than when they proceed from an indexical bond alone. This is what makes deoxyglucose imprints interesting: the farther in we detect icons, the farther away we push the skeptic. However, we did not start believing in shapes only in 1982, when Roger Tootell and his colleagues published their findings. So, if shapes-in-the-brain were not needed to believe in shapes-in-the-world, then colours-in-the-brain are not needed to believe in colours-in-the-world (now).

On the non-Lockean worldview that I am trying to recapture, the colours of our shared surroundings are no more mysterious than the shapes of those surroundings. One can, if one wishes, try to reconstruct colour experience with the materials of a colourless world. But, seeing how the very thing sought is excluded from the start, I think that such an endeavour is doomed to fail. The culprit is not experiential qualities, but rather the disenchanted vision of the world that makes such qualities seem out of place. If toying with the chameleonic gland can disabuse us of that worldview, then that will have justified the device.

References

Aho T (2014) Sense perception: early modern theories. In: Knuuttila S, Sihvola J (eds) Sourcebook for the history of the philosophy of mind: philosophical psychology from Plato to Kant, Studies in the history of philosophy of mind, vol 12. Springer, Dordrecht, pp 81–104

Albertazzi L (2012) Qualitative perceiving. J Conscious Stud 19(11–12):6–31

Allen K (2016) A naïve realist theory of colour. Oxford University Press, Oxford

Aristotle (1993) De anima (trans: Hamlyn DW). Clarendon, Oxford

Armstrong DM (1999) The mind-body problem: an opinionated introduction. Westview Press, Boulder

Balibar E (2013) Identity and difference: John Locke and the invention of consciousness (trans: Montag W). Verso, London

Bayne T (2004) Closing the gap? Some questions for neurophenomenology. Phenomenol Cogn Sci 3(4):349–364

Beuchot M, Deely JN (1995) Common sources for the semiotic of Charles Peirce and John Poinsot. Rev Metaphys 48(3):539–566

Boler JF (2004) Peirce and medieval thought. In: Misak C (ed) The Cambridge companion to Peirce. Cambridge University Press, Cambridge, pp 58–86

Byrne A, Hilbert D (2003) Color realism and color science. Behav Brain Sci 26(1):3–21

Carruthers P (2000) Phenomenal consciousness: a naturalistic theory. Cambridge University Press, Cambridge

Chalmers DJ (2015) Panpsychism and panprotopsychism. In: Alter T, Nagasawa Y (eds) Consciousness in the physical world: essays on Russellian monism. Oxford University Press, Oxford, pp 246–276

Champagne M (2015) Disjunctivism and the ethics of disbelief. Philos Pap 44(2):139–163

Churchland PM (1996) The engine of reason, the seat of the soul: a philosophical journey into the brain. MIT Press, Cambridge

Colapietro VM (2010) Peirce today. Pragmatism Today 1(2):7–24

Crane T, Mellor DH (1990) There is no question of physicalism. Mind 99(394):185–206

Dalcourt GJ (1994) Poinsot and the mental imagery debate. Mod Sch 72(1):1–12

Davidson DH (2001) Subjective, intersubjective, objective. Clarendon Press, Oxford

Deacon TW (2011) Incomplete nature: how mind emerged from matter. W. W. Norton, New York

Deely JN (1982) Antecedents to Peirce's notion of iconic signs. In: Lenhart MD, Herzfeld M (eds) Semiotics 1980. Plenum, New York, pp 109–120

Deely JN (1990) Basics of semiotics. Indiana University Press, Bloomington

Deely JN (1994) Locke's philosophy versus Locke's proposal for semiotic. Am J Semiot 11(3–4):33–37

Deely JN (2001) Four ages of understanding. University of Toronto Press, Toronto

Deely JN (2003) The word "semiotics": formation and origins. Semiotica 146(1–4):1–49

Deely JN (2009) Basics of semiotics, 5th edn. Tartu University Press, Tartu

Dodd J (1995) McDowell and identity theories of truth. Analysis 55(3):160–165

Fodor JA (1983) The modularity of mind: an essay on faculty psychology. MIT Press, Cambridge

Furton EJ (1995) A medieval semiotic: reference and representation in John of St. Thomas' theory of signs. Peter Lang, Bern

Furton EJ (1997) The constitution of the object in Immanuel Kant and John Poinsot. Rev Metaphys 51(1):55–75

Goodman N (1976) Languages of art: an approach to a theory of symbols. Hackett, Indianapolis

Harman G (1990) The intrinsic quality of experience. In: Tomberlin JE (ed) Philosophical perspectives, vol 4. Ridgeview, Atascadero, pp 31–52

Harnad S (2002) Symbol grounding and the origin of language. In: Scheutz M (ed) Computationalism: new directions. MIT Press, Cambridge, pp 143–158

Hilbert D (1987) Color and color perception: a study in anthropomorphic realism. Center for the Study of Language and Information, Stanford

Hookway C (2007) Peirce on icons and cognition. In: Priss U, Polovina S, Hill R (eds) Conceptual structures: knowledge architectures for smart applications. Springer, Berlin, pp 59–68
Huxley TH (1866) Lessons in elementary physiology. MacMillan, London
Jackson F (1982) Epiphenomenal qualia. Philos Q 32(127):127–136
Jackson F (1986) What Mary didn't know. J Philos 83(5):291–295
Jackson F (2004) Postscript on qualia. In: Ludlow P, Nagasawa Y, Stoljar D (eds) There's something about Mary: essays on phenomenal consciousness and Frank Jackson's knowledge argument. MIT Press, Cambridge, pp 417–420
Jacovides M (1999) Locke's resemblance theses. The Philosophical Review 108(4):461–496
Jacovides M (2007) Locke's distinctions between primary and secondary qualities. In: Newman L (ed) The Cambridge companion to Locke's "Essay concerning human understanding". Cambridge University Press, Cambridge, pp 101–129
James W (1975) The meaning of truth. Longmans, Green, and Company, New York
James W (2007) The principles of psychology, vol 2. Cosimo, New York
Johnston M (1992) How to speak of the colours. Philos Stud 68(3):221–263
Kelley D (1988) The evidence of the senses: a realist theory of perception. Louisiana State University Press, Baton Rouge
King P (1994) Scholasticism and the philosophy of mind: the failure of Aristotelian psychology. In: Horowitz T, Janis AI (eds) Scientific failure. Rowman and Littlefield, Lanham, pp 109–138
Kirkham RL (1997) Theories of truth: a critical introduction. MIT Press, Cambridge
Ladyman J, Ross D, Spurrett D, Collier J (2007) Every thing must go: metaphysics naturalized. Oxford University Press, Oxford
Legg C (2013) What is a logical diagram? In: Moktefi A, Shin S-J (eds) Visual reasoning with diagrams. Birkhäuser, Heidelberg, pp 1–18
Levine J (2007) Phenomenal concepts and the materialist constraint. In: Alter T, Walter S (eds) Phenomenal concepts and phenomenal knowledge: new essays on consciousness and physicalism. Oxford University Press, Oxford, pp 145–166
Lewtas P (2013) What it is like to be a quark. J Conscious Stud 20(9–10):39–64
Livingston PM (2004) Philosophical history and the problem of consciousness. Cambridge University Press, Cambridge
Locke J (1825) An essay concerning human understanding. Thomas Davison, London
MacFarlane J (2014) Assessment sensitivity: relative truth and its applications. Oxford University Press, Oxford
Marmo C (1987) The semiotics of John Poinsot. Versus 46:109–129
Maroosis J (2003) Poinsot, Peirce, and Pegis: knowing as a way of being. In: Redpath PA (ed) A Thomistic tapestry. Rodopi, Amsterdam, pp 157–176
McDowell J (1996) Mind and world. Harvard University Press, Cambridge
McGinn C (1989) Can we solve the mind-body problem? Mind 98(891):349–366
Murphy JB (1994) Language, communication, and representation in the semiotic of John Poinsot. Thomist 58(4):569–598
Nagel T (2012) Mind and cosmos: why the materialist neo-Darwinian conception of nature is almost certainly false. Oxford University Press, Oxford
O'Dea J (2002) The indexical nature of sensory concepts. Philos Pap 31(2):169–181
Papineau D (2003) Theories of consciousness. In: Smith Q, Jokic A (eds) Consciousness: new philosophical perspectives. Oxford University Press, Oxford, pp 353–383
Papineau D (2006) Comments on Galen Strawson. In: Strawson G, Freeman A (eds) Consciousness and its place in nature: does physicalism entail panpsychism? Imprint Academic, Exeter, pp 100–109
Papineau D (2007) Phenomenal and perceptual concepts. In: Alter T, Walter S (eds) Phenomenal concepts and phenomenal knowledge: new essays on consciousness and physicalism. Oxford University Press, Oxford, pp 111–144
Peirce CS (1931–58) The collected papers of Charles Sanders Peirce. Harvard University Press, Cambridge

Peirce CS (1998) The essential Peirce: selected philosophical writings, vol 2. Indiana University Press, Bloomington
Peirce CS (2015a) Division of signs. Sign Syst Stud 43(4):651–662
Peirce CS (2015b) Two papers on existential graphs by Charles Peirce. Synthese 192(4):881–922
Pellerey R (1989) Thomas Aquinas: natural semiotics and the epistemological process. In: Eco U, Marmo C (eds) On the medieval theory of signs. John Benjamins, Amsterdam, pp 81–105
Poinsot J (2013) Tractatus de signis: the semiotic of John Poinsot (trans: Deely JN, Powell RA). St. Augustine's Press, South Bend
Potrč M (2008) The world of qualia. In: Wright E (ed) The case for qualia. MIT Press, Cambridge, pp 109–124
Prinz JJ (2002) Furnishing the mind: concepts and their perceptual basis. MIT Press, Cambridge
Pylyshyn ZW (1984) Computation and cognition: toward a foundation for cognitive science. MIT Press, Cambridge
Quine WVO (1966) The ways of paradox and other essays. Random House, New York
Ransdell JM (1978) A misunderstanding of Peirce's phenomenology. Philos Phenomenol Res 38(4):550–553
Ransdell JM (1979) The epistemic function of iconicity in perception. In: Ketner KL, Ransdell JM (eds) Studies in Peirce's semiotic. Institute for studies in pragmaticism, Lubbock, pp 51–66
Ransdell JM (2003) Iconicity revisited. Semio Inq 23(1–3):221–240
Rasmussen DB (1994) The significance for cognitive realism of the thought of John Poinsot. American Catholic Philos Q 68(3):409–424
Rosenthal DM (2010) How to think about mental qualities. Philosophical Issues 20:368–393
Runes DD (ed) (1980) Dictionary of philosophy: ancient, medieval, modern. Littlefield, Totowa
Russell B (2015) My philosophical development. In: Alter T, Nagasawa Y (eds) Consciousness in the physical world: essays on Russellian monism. Oxford University Press, Oxford, pp 46–57
Santaella L (2003) Why there is no crisis of representation, according to Peirce. Semiotica 143(1–4):45–52
Searle JR (1980) Minds, brains, and programs. Behav Brain Sci 3(3):417–424
Sebeok TA, Danesi M (2000) The forms of meaning: modeling systems theory and semiotic analysis. Mouton de Gruyter, Berlin
Sellars W (1991) Science, perception and reality. Ridgeview, Atascadero
Shoemaker S (1994) Self-knowledge and inner-sense; lecture III: the phenomenal character of experience. Philos Phenomenol Res 54(2):291–314
Strawson G (2006) Consciousness and its place in nature: does physicalism entail panpsychism? Imprint Academic, Exeter
Stroud-Drinkwater C (1994) The naive theory of colour. Philos Phenomenol Res 54(2):345–354
Sundström P (2007) Colour and consciousness: untying the metaphysical knot. Philos Stud 136(2):123–165
Tiercelin C (2006) The importance of the medievals in the constitution of Peirce's semeiotic and thought-sign theory. In: Fabbrichesi R, Marietti S (eds) Semiotics and philosophy in Charles Sanders Peirce. Cambridge Scholars, Newcastle upon Tyne, pp 158–184
Tootell RBH, Silverman MS, Switkes E, Valois RLD (1982) Deoxyglucose analysis of retinotopic organization in primate striate cortex. Science 218(4575):902–904
Tye M (1995) Ten problems of consciousness: a representational theory of the phenomenal mind. MIT Press, Cambridge
Tye M (2000) Consciousness, color, and content. MIT Press, Cambridge
Violi P (2007) Semiosis without consciousness? An ontogenetic perspective. Cogn Semiot 1:65–88
Weinberg S (2016) Consciousness in Locke. Oxford University Press, Oxford
Wild J (1947) An introduction to the phenomenology of signs. Philos Phenomenol Res 8(2):217–233
Wittgenstein L (2002) Tractatus logico-philosophicus (trans: Ogden CK, Ramsey FP). Routledge, London

Chapter 6
Informational Flow Implies Informational Pause

Abstract In the previous chapter, I tried to keep ordinary colour perception from being philosophically dismissed. In this chapter, I want to argue that countenancing such qualities at a fundamental level is more promising than waiting for those qualities to emerge at higher levels of complexity. Although complex patterns are crucial to the deployment of neuroscientific and psychological explanations, prescission allows us to artificially decompose those patterns into simple qualities which, owing to their radical isolation from all relations, elude all scientific scrutiny. I argue that, so long as humans have this ability to prescind, worries about phenomenal consciousness will resurface. I thus try to demystify our powers of prescission by examining the Game of Life setting. Philosophers of mind like Daniel Dennett have used this setting to explain intentional explanation, but I think it can also be used to explain how we arrive at the idea of qualia.

6.1 Informational Flow

When it comes time for Chalmers to produce a positive account in *The Conscious Mind*, he turns to information theory for inspiration. I too am drawn to the "strangely beautiful conception" of the world "as pure informational flux, [...] a world of primitive differences, and of causal and dynamic relations among those differences" (Chalmers 1996, 303). Yet, because a tenable philosophy of mind must countenance more than just differences, I agree with Søren Brier that if Chalmers is to successfully deploy a concept of information that is "beyond functionalism and computationalism," he will need "to add Peirce's semiotic philosophy to his theory" (Brier 2008, 38). Let me therefore venture such an addition.

Semiosis is crucial to an information-based account of the mind, because semiosis is how information flows. The neuroscientist Bernard Baars writes that "[i]f we could zoom in on one individual neuron [...] we would see the nerve cell *communicating* frantically to its neighbors about one thousand times per second [...]" (1997, 18, emphasis added). Frederik Stjernfelt (in Emmeche et al. 2008, 7) observes that, promissory notes aside, such appeals to "communication" are never reduced or eliminated. A philosopher of signs would argue that such reductions or eliminations are never carried out because, fundamentally, the action of signs is a genuine part of

M. Champagne, *Consciousness and the Philosophy of Signs*, Studies in the History of Philosophy of Mind 19, https://doi.org/10.1007/978-3-319-73338-8_6

reality. If, say, neuron A impinges directly on neuron B, and neuron B impinges directly on neuron C, then no matter how we unpack the impingement, we cannot credit these two events with any kind of informational transfer unless neuron C is affected in a way that makes it relate to neuron A through the intermediary of neuron B. This is what Peirce meant by semiosis, namely "an action, or influence, which is, or involves, a cooperation of *three* subjects, such as a sign, its object, and its interpretant, this tri-relative influence not being in any way resolvable into actions between pairs" (1998, 411).

Fred Dretske (1981) has done a great job of showing how this flow of information can be used to craft a persuasive account of the mind. Unfortunately, Dretske (1995) also thinks we should try to "naturalize" the mind in a way that makes the idea of qualitative experience dubious (see Bailey 2005). I do not see why this incredulity should follow. To illustrate why I think this, I want to use a cybernetic setting called the Game of Life.

The Game of Life is not a "game" at all, but rather a self-organizing system invented in the 1960s by the mathematician John Conway (see Poundstone 1985, 24). It consists of a primitive set of axioms or "rules" deployed on a two-dimensional grid of cells. In the grid, any given cell will have eight neighbouring cells. Depending on how many of those neighbouring cells are "on" or "off," the cell at the center will be either on or off (normally rendered as black or white). If, for example, two neighbours are on, the center cell will maintain its status (as either on or off) in the next generation or time-slice. Other rules apply to other states. As these rules are successively implemented, the dots begin to have a life of their own. Indeed, viewers looking at the Game of Life can eventually run a rough and ready "zoology" of the patterns. For instance, one can tell that a pattern of dots that looks like a "glider" is about to fall prey to an incoming "eater" (Poundstone 1985, 40).

Although most of these patterns eventually degrade, patterns in a sufficiently complex version of the Game of Life might end up reproducing themselves in an "autopoietic" manner (see the criteria in Thompson 2007, 101–103). For all I know, a pattern might even develop "an interest in conserving its own structure" (Brier 2008, 259; see Champagne 2011). More modestly, we can follow Don Ross and consider a pattern real if it is "projectible [*sic*] under at least one physically possible perspective" and the pattern "cannot be tracked unless the encoding is recovered from the perspective in question" (2000, 161).

I am not the first to be impressed with the patterns that emerge in the Game of Life. Gregg Rosenberg (2004) refers to them a lot, as does Daniel Dennett (1991b). Dennett emphasizes how categorizing the morphological types of the Game of Life gives one some measure of predictive power. According to him, folk psychology works because it latches onto such real patterns. Looking at the behaviour of humans, Dennett argued that, although everything is made of matter, ascribing motives and beliefs is an objective strategy, in that we can robustly gauge whether it renders its predictive service. As he put it: "The decision to adopt the intentional stance is free, but the facts about the success or failure of the stance, were one to adopt it, are perfectly objective" (Dennett 1987, 24). Predicting the behaviour of a toddler is admittedly harder than predicting the behaviour of a coarsely pixelated glider. Still, in either case, the predictions work because they discern real patterns.

This account allows for two perspectives. There is, on the one hand, a base level where all that one countenances are the cell pixels themselves, considered as immanent particulars. Since there is a one-to-one correspondence between this perspective's representations and the discrete cells that are either on or off, the information-theoretic depiction is appropriately called a "bit-map" (literally, a "map of binary digits"). Such a map robustly charts an area with a degree of accuracy proportionate to the number of divided squares. Given an exhaustive statement of the initial conditions of the game—that is, of the (finite) rules and starting positions of the (finite) pieces on the grid—one could in principle deduce (in Laplace-like fashion) the various positions these will occupy on the board as the discrete volleys of moves are repeatedly implemented. The inferential process underwriting such a systematic forecast would be completely monotonic or truth-preserving: working with a bit-map is informationally onerous, but assuming that the governing laws do not change midway, it yields a predictive output that is as secure as its input.

The inferential dynamic at work is radically different, however, when one adopts the intentional stance. Indeed, if one ascends to a higher-level perspective where creature-like actors can be discerned on the playing field, one does so at the price of turning to an inductive rationale that has a far less mechanical character. Dennett is forthright about the fact that a concession along these lines needs to be made in order to have access to intentional explanation. Whereas an exhaustive account of the pixels' dispersion on the grid is a robust affair, moving away from a bit-map and limning "abstracta" (Dennett 1991b, 28) like "eaters" involves a generalization from particulars which effectively soils the truth-preservation with informational noise. This allows one to draw predictions only "sketchily and riskily" (Dennett 1991b, 40). A considerable boon of yielding to such induction, however, is that it provides one with assorted heuristics that are unavailable when one countenances only the pixels themselves.

To bring out the difference between these two perspectives, we might imagine having to text message another person's demeanour and location with a mobile phone. If one were to take the bit-map route and catalogue the position and velocity of all the person's constituent particles, this would make for a lengthy message (and run a costly bill). In information-theoretic parlance, the message would have greater entropy. But, if one were to type something along the lines of "So and so is sitting in such and such a room doing this or that," one would thereby spare oneself a lot of bits (not to mention a lot of money). Thanks to this, the configuration of all the relevant neurons in a given brain state need not be exhaustively recounted to get a sense of what the subject is thinking/feeling. Neuroscientific explanation is not impotent, but in comparison with the intentional stance, it makes a very long detour.

In principle, the informational compressibility involved in folk psychology—no matter how cognitively cost-effective—should be insufficient to sway a materialist into countenancing things (like "eaters" or "angry people") whose ontology is, at root, "more" than merely material. A materialist would try to enumerate the individual pixels of the Game of Life which a third party espousing the intentional stance has the liberty to coalesce and taxonomize. True, "posing fanciful interpretations" (Dennett 1991b, 41) spares one the tedious task of having to describe the domain in question one atomic unit at a time, but a materialist might reply that a token-only approach is the surest way to ensure that one's account hugs the metaphysical makeup of the world as closely as possible. What sets materialism apart from the intentional stance, then, is that it does not give any weight to considerations of cognitive economy. If ease of use mattered as much as matter, materialism would not have to wait for a completed neuroscience to tell it what anger really is.

Informational compression is asymmetrical in that it can result in a loss of data (especially if the domain represented is disorderly). Going back to the example just used, from an exhaustive account of another person's material makeup and spatial location, one can infer "who" and "where" (colloquially) they are; but from simply being told that "So and so is sitting in such and such a room doing this or that," one cannot determine the person's exact spatial coordinates. That is the price to pay for engaging in intentional talk: we save time precisely because we cut corners. The downside is that there is just no way to unpack all the observational consequences of a given intentional statement, so in this respect the idiom will always fall short of a "complete" scientific account. I want to argue, however, that informational compression is something no science can do without.

Bit-mapping the world is a chimerical aspiration. To take an example from astronomy, we do not exhaustively chart the path and constitution of celestial bodies, pile up the ensuing records in some museum vault, and call that knowledge. Rather, "[t]he positions of the planets in the solar system over some interval constitute a compressible data set, because Newton's laws [...] supply the necessary algorithm to achieve the compression" (Davies 1990, 63). This informational compression is required at small scales too. Even talk of particles is shorthand for talk of patterns. As Peirce rightly noted,

> The existence of things consists in their regular behaviour. If an atom had no regular attractions and repulsions, if its mass was at one instant nothing, at another a ton, at another a negative quantity, if its motion instead of being continuous, consisted in a series of leaps from one place to another without passing through any intervening places, and if there were no definite relations between its different positions, velocities and directions of displacement, if it were at one time in one place and at another time in a dozen, such a disjointed plurality of phenomena would not make up any existing thing. (Peirce 1992, 278)

According to Peirce, "substances" are in fact "constituted by regularities" (ibid.). As far as explanations go, the antics of one's conspecifics are no different. People do things in a patterned way, and it is the business of psychology and cognitive science to discern those patterns in order to predict what people will do next.

To miss a pattern that is potentially visible is surely to miss relations that truly exist in the world. Observers nevertheless privilege some patterns over others because they bring their aims, queries, and practical interests to bear on the transaction (Peschard 2010). Using an example from Henry James, we can say that making out a complex figure in a Persian carpet is a joint effort, the subject selecting one shape among others, the object ensuring that this selection is not confabulated.

Hence, in this account, "any act of interpretation is a dialectic between [...] initiative on the part of the interpreter and contextual pressure" (Eco 1990, 21).

6.2 Informational Pause

The informational account I have just sketched uses patterns to explain familiar things and events like tables, frogs, jealousy, good jokes, and economic recessions. Is this enough? Don Ross thinks so. In fact, he believes that "reality is composed of real patterns *all the way down*" (Ross 2000, 160; emphasis in original). I want to part ways with Ross here because, if patterns are complex and complexity subsumes simplicity, then those patterns must subsume simple qualities. To see this, we can look again at the Game of Life setting and make the following observations:

1. It is complex.
2. There is no such thing as a neighbourless cell pixel.

Now, consider what happens when we add the following claim:

3. Complexity subsumes simplicity.

I hold these three claims to be true. Yet, their conjunction can create a tension because, in principle, claims (1) and (3) allow for the supposition of a neighbourless cell pixel—even though claim (2) states that, factually, there is no such thing. So long as humans are capable of realizing this, the worries that generate the "hard problem" of consciousness will persist. By inserting a formal distinction in the subsumption of (3), prescission lets us see how the conjunction of these claims can be consistent.

Like the Game of Life, the field of human experience conjoins complexity and simplicity. However, Bertrand Russell held that science has nothing to say about simple qualities (Holman 2008, 50–51). To arrive at this view, Russell implicitly relied on a premise which, following Stathis Psillos (2009, 126), we may call the "Helmholtz-Weyl" principle. It states that "we are justified, when different perceptions offer themselves to us, to infer that the underlying real conditions are different" (quoted by Weyl 1963, 26). This principle grafts itself nicely onto other Russellian tenets. As we saw in Chap. 4, Russell (1998, 59) thought that observational episodes conveyed by nondescript demonstratives like "This is white" supply us with "knowledge by acquaintance." Although we can attempt to convey by ostension what is happening when we enjoy first-person episodes, these empirical points of contact, Russell held, are by their nature private and ineffable. This does not bode well for objective third-person knowledge. Not to worry, one does not have to wait long to depart from this, since "[t]he next simplest [facts] would be those in which you have a relation between two facts, such as: 'This is to the left of that'" (Russell 1998, 59). The moment we bring two or more relata into relation, we leave the domain of intrinsic qualities and effectively enter that of intelligibility (and, by extension, science).

This is where the move from different perceptions to different causes does its work. On this view, a three-part sequence like lemon/apple/lemon will bear the same relational configuration as the three-part sequence avocado/banana/avocado—even if one person tastes avocado flavours where another tastes lemon. "Two relations *P*, *Q* are said to be 'similar' if there is a one-one relation between the terms of their fields, which is such that, whenever two terms have the relation *P*, their correlates have the relation *Q*, and vice versa" (Russell 1954, 249). In the previous example, avocado-tastes mapped onto lemon-tastes, and apple-tastes onto banana-tastes. Likewise, "[a] book spelt phonetically is similar to the sounds produced when it is read aloud" (Russell 1954, 249), since the structure binding the printed characters can be monotonically correlated with the structure binding the spoken sounds (ibid., 400). Hence, so long as variations in experience attest to variations in whatever is impinging on the sense organs, the Helmholtz-Weyl principle licenses the inference of a common structure. "In this mathematical view, structure is a domain of similarity and difference, which, like color for the blind, has no substance of its own" (Lidov 1999, 128).

Moritz Schlick, the leader of the Vienna Circle, was influenced by Russell's analysis. In a series of lectures titled "Form and Content," Schlick argued that "in order for *understanding* actually to occur, it would be necessary for the 'structure' or 'form' of linguistic signs to be 'filled in' with 'content' drawn from individual experience" and that this filling in is what "allows empirical propositions *ultimately* to be justified by experience" (Livingston 2013, 82). Schlick therefore held that qualities would have to be included in a full account because, "if we are to have a science of some domain of reality instead of a mere hypothetical-deductive system, then our symbols must stand for real content" and not "mere structure" (1979, 331). As a result, Schlick's stance on Frank Jackson's knowledge argument would be that "one can learn the meaning of the words 'joy' or 'green' only by being joyful or seeing green" (ibid., 321).

What is it like to sip a cappuccino? Using prescission, tone, and iconicity, Peircean philosophy of signs can demystify the fact that, "If you got to ask, you ain't never gonna get to know" (Louis Armstrong, quoted in Block 1978, 281). However, because they were using Russellian tools instead of Peircan tools, Schlick's analytic colleagues were unsure what to make of this "inexpressibility of content" (1979, 300). Rudolf Carnap, for instance, allowed "indicator signs" (Kennzeichungen) to track things, but the question of "what" those indices track did not concern him. Quoting Henri Poincaré approvingly, Carnap held that "only the relations between the sensations have an objective value" (2003, 30). Russell also conceded that the qualitative contents which hang together in a given structure fall outside the ambit of testability and inter-subjective verification. According to Russell, when two systems of relations are isomorphic, "the only difference must lie in just that essence of individuality which always eludes words and baffles description [...]" (Russell 1950, 61).

Wittgenstein agreed with Russell, Schlick, and Carnap that linguistic structures leave something out. Logical analysis can push all the way to single qualities, but Wittgenstein (2002, 189) was more open than his colleagues about the need to shut up when we get to that point. In so doing, Wittgenstein allied himself with "numerous arguments for humility about the intrinsic nature of our world" (Majeed 2013, 259). Even so, by intimating iconicity, "the early Wittgenstein saw further than many of his contemporaries here, drawing his famous distinction between what can be 'said' and what can only be 'shown'" (Legg 2008, 214).

As these stances show, the early analytic tradition struggled with the same themes and issues that currently animate philosophical debates about consciousness. It is bizarre, then, to see Paul Churchland dismissing arguments that decompose complex relations into qualitative simplicity as "lack[ing] integrity even by the standards of purely analytic philosophy" (2011, 18). If Churchland wants to say that the function/quality distinction is bogus, then he must explain why so many brilliant thinkers have been (and continue to be) drawn to it.

Philosophers are not the only ones to have sensed that functional descriptions fail to tell the whole story. The British astronomer Sir Arthur Eddington noted that the knowledge we have of particles comes from, "like everything else in physics, a schedule of pointer readings [on instrument dials]" (quoted in Strawson 2006, 10). These patterns of pointer readings permit an indexical tracking of whatever causally interacts with a given instrument, but the question of what this tracked object is remains unaddressed. Ned Block illustrates this with a clever argument. He begins by noting that particles are individuated in physics by "having certain lawlike relations to certain other physical properties" (Block 1978, 302). Relations like these allow scientists to pick out "dual particles" like protons and anti-protons. Such particles are said to be "dual" because their relations are identical. Physicists nevertheless distinguish between the two kinds of particles because, when they are combined, they annihilate each other. So, even though "physics characterizes its basic entities only *extrinsically*, in terms of their relations to other entities" (Chalmers 2010, 27), the assumption that particles possess intrinsic natures is all that prevents what we call protons from being called anti-protons and vice versa (Block 1978, 302).

Some philosophers of science (notably Worrall 1989) have claimed that focusing on what a thing "does" and staying mum on what a thing "is" allows us to find theoretical continuities that survive paradigm shifts. The silence called for by this "structural realism" offends some conceptions of science (Psillos 2009). However, I think that, in philosophy of mind, a satisfactory account should leave us able to distinguish lemon/apple/lemon from avocado/banana/avocado.

How should we understand the intrinsic qualities of things apart from the sundry relations they entertain? For Eddington, the answer rests on an inference to the best explanation: manifestly, conscious experience lets us appreciate what qualities are. Since we have no other or better way to apprehend the intrinsic character of a thing, "it seems rather silly to prefer to attach [a schedule of pointer readings] to something of a so-called 'concrete' nature inconsistent with thought, and then to wonder where the thought comes from" (quoted in Strawson 2006, 10). This is similar to Schlick (1979, 296), who said that we must "fill-in" structure with qualia. True, if we consider a quality in a way that abstracts away all relations, we are left with something that "is irrelevant to science" (Russell 1950, 61). It takes an added premise of scientism, however, to infer that irrelevance to science means irrelevance to our lives.

Strawson does not endorse scientism, and neither do I. In contrast with Strawson, though, I do not think that the simple qualities obtained by prescission entail any kind of panpsychism. Although it is common to assume that "whatever does not belong to the structure [...] is, in the final analysis, subjective" (Carnap 2003, 29), this is a Lockean assumption that I reject. Like Peirce, I believe that, when we arrive at a quality by stripping away all relations (e.g., patterns, structures, functional roles, etc.), we are no longer entitled to locate that quality in the mind since, in principle, there should be no way to tell. Eddington and Strawson think that physical science "allows that among such intrinsic properties could be phenomenal properties" (Holman 2008, 53). I agree with Eddington and Strawson that we can have a direct appreciation of such qualities, but I see no reason why this direct appreciation should require introspection. Iconicity is a sign-theoretic way to evince a direct appreciation of qualities without foundering into any kind of world-wide psychologism.

The individual cells in the Game of Life are not empty. Acknowledging this is, I think, more sensible than compounding pattern upon pattern in the hope that qualitative experiences will eventually emerge. Hence, "[i]n Peirce's semiotics, everything in nature is a potential sign. [...] The implication of this is that qualia, and 'the inner life' are potentially there from the beginning. [...] The point is that organisms and their nervous systems do not create mind and qualia" (Brier 2008, 99). This account squares with our actual experience. Discussing consciousness, the semiotician David Lidov writes:

> What, [Dennett] asks, was the taste of beer the first time you tasted it, before you learned to like it? He seems to think that it had no specific taste, no *qualia*. Dennett's understanding seems to be that what we mistook for taste was really a frustrated urge to spit the stuff out. Such a disposition may well have been part of the experience, but I also remember a taste. (Lidov 1999, 117)

According to my philosophy of signs, Lidov is not blundering when he thinks that the beer was bitter. Responses to beer may create a pattern, but the quality that those patterned responses respond to can be isolated using prescission.

6.3 Can a Pragmatist Accept this?

I have just told a story about how experiential qualities can be supposed apart from their detectable influence on other things. Can a pragmatist committed to tangible effects accept this? I don't see why not. As the "Ur-pragmatist" (McDermid 2008, 2), Peirce did say, quite rightly, that if we want to make our ideas clear(er), then we should look downstream to their actual and future effects (Peirce 1992, 132). Yet, there is a tendency to overlook that this pragmatist recommendation is nested in a conditional: *if* you want clarity, then you should do this and that. I see no reason—and, to my knowledge, Peirce gave no reason—why anyone should feel obliged, in the strict deontological sense, to pursue intellectual clarity, come what may. Rather, that pursuit, which finds its maximal expression in collective scientific inquiry, seems to be one among many.

In "The Fixation of Belief" (1992, 109–123), Peirce uses a series of foils to progressively build up to the view that he finally wants us to endorse. One method of settling opinion is presented charitably, then a flaw is detected, which leads to the development of a better method, followed by another flaw, and so on—until we reach the fourth and final method, that of science. By pooling our results and keeping our judgements open to revision, the scientific method lets us turn our fallible shortcomings into learning opportunities. Of course, it is normal to think that, if you are offered various options and you know which is best, then you should pick that best one. So, predictably, when Peirce offers us three "grades of clearness" culminating in his pragmatic maxim (1992, 132), we naturally assume that the other two were there mainly to rhetorically set the stage. However, the moment we do this, we walk away from the possibility of something not defined by its causal or inferential role(s).

William James was ready to relax the demands of scientific inquiry whenever following those demands would result in increased personal suffering. John Dewey expressed similar humanistic concerns while putting greater emphasis on collective benefit as the bottom line. Yet, in their rush to be branded as naturalists, many pragmatists have lost sight of the fact that not everything is appropriately gauged by its current or anticipated practicality. To borrow an example favoured by Ned Block, "there are features of the experience of orgasm that don't represent anything" (1995, 34). So, when you enjoy one of those, your aim cannot possibly be the end of inquiry. I therefore believe pragmatism becomes more plausible when it makes room for inefficiency—in the double sense of a respite from technological progress and an escape from efficient causation.

In the philosophical literature on concepts, it is common to ask whether concepts are structured or unstructured (Margolis and Laurence 1999, 4–5). Whether or not it is appropriate to call it a concept (see Prinz 2007), the simple quality reached by prescissive abstraction is clearly unstructured. Language requires structure (think of grammar, for instance). Hence, there is not much one can say about a quale. Even so, these tones are not completely useless, insofar as their very quality can prompt/support at least one anticipation, namely that something else like it might show up again in our stream of consciousness.

This free inference that iconicity affords is as epistemically weak as it gets—we might even call it "the platitudinous affordance." Derk Pereboom (2011) has explored the possibility that one might represent qualitative natures that the objects of those representations do not in fact have. That is certainly a possibility. However, while one can doubt whether a hypothesis is true, one can hardly doubt that a hypothesis is a hypothesis. The stream of consciousness suggests that something might hold, but the lack of a tangible verdict muddles the distinction between veridical and illusory experience, such that "the sorting out of which-is-which is a problem rather than a given" (Deely 2003, 188). That is why, in actual (i.e., unpaused) semiosis, we have to act in order to perceive (Clark 2013). So, while Peirce's pragmatism shows "How to Make Our Ideas Clear" (1992, 124–141), his philosophy of signs shows why our ideas need to be made clear in the first place.

6.4 Classifying the Peircean Stance

Admittedly, the semiotic account of consciousness that I have been developing in this book does not fit well within the popular rivalry between (materialist) monism and dualism. Taking that rivalry for granted, Chalmers (2010, 111–137) has proposed a six-fold division of metaphysical stances. His first trio of views (A to C) shares a commitment to materialism. Type-A materialism denies that we face any explanatory gap when it comes to the mind. Type-B materialism accepts the existence of such a gap in knowledge, but denies that it reflects an ontological gap. Type-C materialism admits that there is presently an epistemic gap, but holds that it will eventually be closed. Pausing mid-way, then, Daniel Dennett (1991a) would be an advocate of type-A, Brian Loar (1997) of type-B, and Paul Churchland (1981) of type-C.

The metaphysical stances of Chalmers' second trio (D to F) all countenance more than matter. Type-D dualism holds that the mind not only exists, but is causally efficacious as well. Type-E epiphenomenalism grants the separate existence of the mind, but denies its causal efficacy. Finally, type-F countenances the mind as a fundamental ingredient, but locates it at such a basic level that even matter can be said to be minded (or proto-minded). Popper and Eccles (1981) argue for type-D, Frank Jackson (1982) once defended type-E, and Galen Strawson (2006) currently advocates type-F.

Let us take a closer look at type-F views. A position falls under the "panpsychist" genus when it attempts to side-step the problem of emergence by making consciousness or mind a basic metaphysical ingredient. As the subtitle to his book *The Conscious Mind* attests, Chalmers is in search of a "fundamental" theory of consciousness, where fundamental features are those that "cannot be explained in terms of more basic features" (1996, 126). In his sequel, *The Character of Consciousness*, Chalmers expresses his conviction that "in some ways the type-F view is the most appealing" (2010, 138).

In much the same spirit as the yearly Toward a Science of Consciousness conferences (often held in Tuscon), semioticians are currently trying to craft a worldview in which meaning does not seem out of place (Deely 1994, 2001; Santaella 2009). In October 2008, the Semiotic Society of America devoted its first ever session to discussing the possible existence of sign-action in the inorganic realm. For reasons that I outline in Champagne (2013), I do not think this project has yet succeeded. I nevertheless agree with panpsychists that countenancing qualities as primitive is "less daunting than articulating a comprehensible theory of radical emergence of mind from utterly mindless matter" (Skrbina 2006, 156). Our shared argument, in effect, is this:

1. It is undeniable that we experience qualities.
2. Qualities do not figure in a materialist ontology.

Therefore,

3. Either a materialist ontology is mistaken, or a materialist ontology must give an account of how our experience of qualities emerges from matter.
4. Nothing comes from nothing.

Therefore (in light of all of the above claims),

5. It is impossible to give an account of how our experience of qualities emerges from matter.

Therefore (in light of premises 3 and 5),

6. A materialist ontology is mistaken.

Therefore,

7. Qualities need to figure in an ontology.

This is my formulation, but the line of reasoning that it captures is not new; it was stated, for example, by Nagel (1979, 181). For the most part, I consider this argument to be sound. Where I differ, though, is that I do not think that accepting the conclusion (7) also entails accepting panpsychism, which we might graft as a supplementary conclusion (8). As such, I am trying to give a plausible alternative story for those who find the argument above persuasive yet who recoil from the prospect of scattering mind everywhere.

One could presumably object to the argument in a couple of spots. One could, for example, reject its starting claim that we experience qualities. After all, if we do not even experience qualities, then the fact that qualities do not figure in a materialist ontology (premise 2) does not pose much of a problem. However, I accept premise (1), because I agree with Strawson that a "[f]ull recognition of the reality of experience [...] is the obligatory starting point for any remotely realistic (indeed any non-self-defeating) theory of what there is" (2006, 4). Premise (5) might also be weakened to something like "No account of the emergence of qualia from matter has thus far been successful." Instead of betting that science will (one day) vindicate eliminativism (Churchland 1981), one could bet that science will (one day) vindicate emergentism (Park 2013 does just that). This would conceivably leave room for optimism, thereby blocking the deductive inference to claim (6). However, I accept premise (5) in an undiluted form, because I think the burden is on whoever holds it to be false to show why it is not true.

Panpsychism certainly has a respectable historical pedigree. David Skrbina points out that "many of our greatest thinkers and philosophers have held to some version of panpsychism" (2006, 152) and he includes Peirce in that list. There is textual justification for this inclusion. Peirce wrote, for example, that "matter is effete mind, inveterate habits becoming laws" (1992, 293). Yet, in another place, he also wrote that the "universe is perfused with signs, if it is not composed exclusively of signs" (Peirce 1998, 394). That is not exactly panpsychism. Since every triadic sign has an element of Firstness, and since nothing in Firstness specifies whether it is mental or material, it might be more appropriate to call Peirce's commitment "panqualityism" (Chalmers 2015, 270–274; Coleman 2015). I nevertheless want to forgo that label, since panqualityism would constitute only one third of a semiotic universe. As a result, "[h]owever tempted Peirce may have been by a panpsychist approach to semiotic, [...] such a position is rendered impossible by his own principles of semiosis" (Kruse 1990, 222).

Panpsychists can certainly find support for their view in some Peircean passages (Skrbina 2005, 152–155). However, my sense is that most philosophers of mind who turn to panpsychism are not so much interested in panpsychism, but rather in the idea of putting qualities back into the world—panpsychism just happens to be the only stance they know that does this (see Goff 2017). This may explain why the "psyche" portion of the stance is often weakened to such an extent that it is barely recognizable as mind. Given this weakening, there seems to be little sense in considering a stance like panqualityism to be a species of the genus panpsychism. What could be the common trait here? Surely the "pan" is not sufficient for membership in that genus. If it were, then materialism would have to also be included, since materialism, rigorously stated, is really panmaterialism.

Firstness is fundamental, so it can make Peircean semiotics look like a type-F view. However, because Peircean semiotics also countenances Secondness and Thirdness, the stance does not fit well in Chalmers' classification. Like Chalmers, Peirce (1998, 179–195) also classified metaphysical positions, depending on which of his three categories they are ontologically committed to. Exhausting the combinations left him with seven possible options. Here are Peirce's (somewhat antiquated) labels:

- Nihilism or Idealistic Sensualism, which grants reality to Firstness only.
- Strict Individualism, which grants the reality of Secondness only.
- Hegelianism, which grants the reality of Thirdness only.
- Cartesianism, which grants the reality of Secondness and Thirdness.
- Berkeleyanism, which grants the reality of Firstness and Thirdness.
- Ordinary Nominalism, which grants the reality of Firstness and Secondness.

And, finally,

- "The metaphysics that recognizes all the categories" (Peirce 1998, 180).

Because the three categories of semiotic analysis are cardinal and not ordinal, I wish Peirce had said "...recognizes all the categories *and their ordinal arrangement*." At any rate, I find his classification more helpful. A triple-layer account inspired by Peirce—coupled with a better understanding of our prescissive powers to asymmetrically suppose some of those layers absent—can inspire an ontology well-suited for those suspicious of the claim that reality comprises something non-structural (see Champagne 2015).

If iconicity can let us see the world as it is and if what we see includes qualities like colours (as I argued in Chap. 5), then we have good reason to think that the world includes qualities like colours. However (as I argued in Chap. 3), we have no reason whatsoever to think that such qualities can be conceived apart from our prescissive intervention. We must therefore conjoin two seemingly contradictory theses: 1) prescission is something we (humans) do, and 2) it is not up to us (humans) what results when prescission is properly carried out. Since the first thesis speaks to mind-dependence and the second thesis speaks to mind-independence, their conjunction can seem unstable because, usually, something is considered real only when it is mind-independent. It would of course simplify matters if the interpretations and objects of signs could really *be* absent, not just *supposed* absent. I think, though, that we have much to gain from getting used to this nuance.

Some complex patterns are pattern-grasping. When these pattern-grasping patterns are human creatures, what "emerges" is not a qualitative dimension that other things lack, but a unique power to isolate the individual qualities that comprise patterns. Wedging formal distinctions into semiosis gives us, not just relations (Bains 2006), but relations and relata. The relations need relata, but the relata do not need relations. Indeed, "[i]t is frequently the case, that, while *A* cannot be prescinded from *B*, *B* can be prescinded from *A*" (Peirce 1992, 3). This asymmetry means that we will never be able to give a scientific description of the intrinsic qualities that fall under Firstness. But, that is simply because no Game of Life (and no language game) can occur with just one cell (or just one quality).

Of course, nothing I have said in this book shows that qualia can be encountered apart from their practical effects. Asking for an example of qualitative Firstness unaccompanied by its indexical and symbolic means of detection would be tantamount to "asking a flashlight in a dark room to search around for something that does not have any light shining upon it" (Jaynes 2000, 23). Yet, as inseparable as phenomenal qualia and their practical effects are, I do not see what can stop rational animals from inserting a formal distinction between the two, nor do I see why such a prescissive power should be taken to threaten the normal conduct and valued standing of experimental science. A better understanding of our ability to contemplate unrelated qualities can nevertheless explain why we humans can "form the idea of phenomena that we do not know how to detect" (Nagel 1986, 24).

References

Baars BJ (1997) In the theater of consciousness: the workspace of the mind. Oxford University Press, Oxford

Bailey AR (2005) What is it like to see a bat? Dretske on qualia. Disputatio 1(18):151–177

Bains P (2006) The primacy of semiosis: an ontology of relations. University of Toronto Press, Toronto

Block N (1978) Troubles with functionalism. In: Savage CW (ed) Minnesota studies in philosophy of science, vol 9. Minnesota University Press, Minneapolis, pp 261–325

Block N (1995) Mental paint and mental latex. In: Villanueva E (ed) Philosophical issues, vol 7. Ridgeview, Atascadero, pp 19–49

Brier S (2008) Cybersemiotics: why information is not enough. University of Toronto Press, Toronto
Carnap R (2003) The logical structure of the world (trans: George RA). Open Court, Chicago
Chalmers DJ (1996) The conscious mind: in search of a fundamental theory. Oxford University Press, Oxford
Chalmers DJ (2010) The character of consciousness. Oxford University Press, Oxford
Chalmers DJ (2015) Panpsychism and panprotopsychism. In: Alter T, Nagasawa Y (eds) Consciousness in the physical world: essays on Russellian monism. Oxford University Press, Oxford, pp 246–276
Champagne M (2011) Axiomatizing umwelt normativity. Sign Syst Stud 39(1):9–59
Champagne M (2013) A necessary condition for proof of abiotic semiosis. Semiotica 197:283–287
Champagne M (2015) A less simplistic metaphysics: Peirce's layered theory of meaning as a layered theory of being. Sign Syst Stud 43(4):523–552
Churchland PM (1981) Eliminative materialism and the propositional attitudes. J Philos 78(2):67–90
Churchland PM (2011) Consciousness and the introspection of "qualitative simples". Eidos 15:12–47
Clark A (2013) Whatever next? Predictive brains, situated agents, and the future of cognitive science. Behav Brain Sci 36(3):181–253
Coleman S (2015) Neuro-cosmology. In: Coates P, Coleman S (eds) Phenomenal qualities: sense, perception, and consciousness. Oxford University Press, Oxford, pp 66–102
Davies PCW (1990) Why is the physical world so comprehensible? In: Zurek WH (ed) Complexity, entropy and the physics of information. Westview Press, Boulder, pp 61–70
Deely JN (1994) The grand vision. Trans Charles S. Peirce Soc 30(2):371–400
Deely JN (2001) Physiosemiosis in the semiotic spiral: a play of musement. Sign Syst Stud 29(1):27–48
Deely JN (2003) The impact on philosophy of semiotics. St. Augustine's Press, South Bend
Dennett DC (1987) The intentional stance. MIT Press, Cambridge
Dennett DC (1991a) Consciousness explained. Little, Brown and Company, New York
Dennett DC (1991b) Real patterns. J Philos 88(1):27–51
Dretske FI (1981) Knowledge and the flow of information. Center for the Study of Language and Information, Chicago
Dretske FI (1995) Naturalizing the mind. MIT Press, Cambridge
Eco U (1990) The limits of interpretation. Indiana University Press, Bloomington
Emmeche C, Hoffmeyer J, Kull K, Markoš A, Stjernfelt F (2008) The IASS roundtable on biosemiotics: a discussion with some founders of the field. Am J Semiot 24(1–3):1–21
Goff P (2017) Consciousness and fundamental reality. Oxford University Press, Oxford
Holman EL (2008) Panpsychism, physicalism, neutral monism and the Russellian theory of mind. J Conscious Stud 15(5):48–67
Jackson F (1982) Epiphenomenal qualia. Philos Q 32(127):127–136
Jaynes J (2000) The origin of consciousness in the breakdown of the bicameral mind. Mariner, Boston
Kruse FE (1990) Nature and semiosis. Trans Charles S Peirce Soc 26(2):211–224
Legg C (2008) The problem of the essential icon. Am Philos Q 45(3):207–232
Lidov D (1999) Elements of semiotics. MacMillan, London
Livingston PM (2013) Phenomenal concepts and the problem of acquaintance. J Conscious Stud 20(5–6):71–92
Loar B (1997) Phenomenal states. In: Block N, Flanagan O, Güzeldere G (eds) The nature of consciousness: philosophical debates. MIT Press, Cambridge, pp 597–616
Majeed R (2013) Pleading ignorance in response to experiential primitivism. Philos Stud 163(1):251–269
Margolis E, Laurence S (1999) Concepts and cognitive science. In: Margolis E, Laurence S (eds) Concepts: core readings. MIT Press, Cambridge, pp 3–81

McDermid D (2008) The varieties of pragmatism: truth, realism, and knowledge from James to Rorty. Continuum, London
Nagel T (1979) Mortal questions. Cambridge University Press, Cambridge
Nagel T (1986) The view from nowhere. Oxford University Press, Oxford
Park JJ (2013) The hard problem of consciousness and the progressivism of scientific explanation. J Conscious Stud 20(9–10):90–110
Peirce CS (1992) The essential Peirce: selected philosophical writings, vol 1. Indiana University Press, Bloomington
Peirce CS (1998) The essential Peirce: selected philosophical writings, vol 2. Indiana University Press, Bloomington
Pereboom D (2011) Consciousness and the prospects of physicalism. Oxford University Press, Oxford
Peschard I (2010) Non-passivity of perceptual experience. Contemporary Pragmatism 7:149–164
Popper KR, Eccles JC (1981) The self and its brain. Springer, Berlin
Poundstone W (1985) The recursive universe: cosmic complexity and the limits of scientific knowledge. William Morrow, New York
Prinz JJ (2007) Mental pointing: phenomenal knowledge without concepts. J Conscious Stud 14(9–10):184–211
Psillos S (2009) Knowing the structure of nature: essays on realism and explanation. Palgrave Macmillan, Hampshire
Rosenberg G (2004) A place for consciousness: probing the deep structure of the natural world. Oxford University Press, New York
Ross D (2000) Rainforest realism: a Dennettian theory of existence. In: Ross D, Brook A, Thompson D (eds) Dennett's philosophy: a comprehensive assessment. MIT Press, Cambridge, pp 147–168
Russell B (1950) Introduction to mathematical philosophy. George Allen and Unwin, London
Russell B (1954) The analysis of matter. Dover, New York
Russell B (1998) The philosophy of logical atomism. Open Court, Chicago
Santaella L (2009) Pervasive semiosis. Trans Charles S Peirce Soc 45(3):261–272
Schlick M (1979) Philosophical papers, vol 2. Riedel, Dordrecht
Skrbina D (2005) Panpsychism in the west. MIT Press, Cambridge
Skrbina D (2006) Realistic panpsychism. In: Strawson G, Freeman A (eds) Consciousness and its place in nature: does physicalism entail panpsychism? Imprint Academic, Exeter, pp 151–157
Strawson G (2006) Consciousness and its place in nature: does physicalism entail panpsychism? Imprint Academic, Exeter
Thompson E (2007) Mind in life: biology, phenomenology, and the sciences of mind. Harvard University Press, Cambridge
Weyl H (1963) Philosophy of mathematics and natural science. Atheneum, New York
Wittgenstein L (2002) Tractatus logico-philosophicus (trans: Ogden CK, Ramsey FP). Routledge, London
Worrall J (1989) Structural realism: the best of both worlds? Dialectica 43(1–2):99–124

Chapter 7
Conclusion

Abstract Six chapters ago, I argued that philosophical worries about the qualitative dimension of consciousness should not be taken so seriously that they trigger a scientific search for qualia, but neither should those worries be belittled or dismissed. Throughout the book, I drew on the ideas of C. S. Peirce—specifically his ideas about prescission, tone, and iconicity—to articulate why/how those philosophical worries about consciousness make sense. In this concluding chapter, I summarize the resulting stance by clarifying what it can and cannot do. Philosophy of signs can show that the idea of qualia is licensed by reason. However, it cannot promise to capture such qualia by scientific means. Although artificially inspecting the incipient substructure of sign-action is in many ways a useless ability, it satisfies a distinctively human sense of curiosity and wonder.

It is often thought that consciousness has a qualitative dimension that cannot be tracked by science. Recently, some philosophers have argued that this worry stems from the special nature of the concepts that we use to refer to our conscious states. I have drawn on the neglected branch of philosophy of signs or semiotics to develop a new take on this strategy. Daniel Dennett is right that "'qualia' is a bit of philosophers' jargon" (1998, 141). However, with some exceptions (Crane 2000; Livingston 2004), few have bothered to retrace where that jargon came from. Retracing those origins brings us to C. S. Peirce—and to an entire conceptual apparatus that makes qualia seem far more sensible.

I therefore started by saying that, if we could see how this book's two opening quotes by Peirce are consistent, we would make progress in philosophy of mind. Let us now take stock of what we have done and see if/how those two claims interlock.

The first quote is, in essence, a slogan for a strong functionalist program. It demands that explanations of the mind start with an input that enters "at the gate of perception" and end with an output that exists "at the gate of purposive action" (Peirce 1998, 241). The goal is to construct a plausible theory of what happens in between. Yet, no matter what such a theory might look like, it seems destined to relate states. This involvement of relations would be benign, were it not for the fact that a prevalent construal glosses consciousness as having a non-relational element to it. Raw feels have an intrinsic quality, irrespective of the objects or behaviours

M. Champagne, *Consciousness and the Philosophy of Signs*, Studies in the History of Philosophy of Mind 19, https://doi.org/10.1007/978-3-319-73338-8_7

they are associated with. This is expressed in the second epigraph, which says that each qualitative state "is in itself what it is for itself, without reference to any other" (Peirce 1931–58, 6.224). A tension thus ensues. The functional program calls for an explanation of mental facts in terms of relations, but at least one dimension of our conscious lives seems to involve simple experiential qualities conceived apart from any relation(s).

Qualia are not supposed to enter into any kind of relation with the world or other mental states, otherwise they could in principle be detected, either through their causal efficacy or their participation in inferences. So, if one accepts both opening quotes, one will run into trouble, because the moment an organism acts on the basis of a feeling, this will generate a worry that we are studying that feeling's discernible effects, not the quality of the feeling itself.

How are we to make sense of this? The key, I suggest, is to see how humans can insert distinctions between features that are always bound. So, on my diagnosis, zombie-style arguments proceed too quickly—or, at any rate, they do not pay sufficient philosophical attention to a crucial step. Those arguments ask us to delete qualia from an organism, but when we look more carefully at just what a quale is, we see that this notion involves a deletion of its own. I have made this prior removal of relations my main focus. David Chalmers speaks of "the double life of mental terms" (1996, 16) and emphasizes that while "[o]ur everyday concept of pain presumably combines the two [concepts of phenomenal pain and psychological pain] in some subtle weighed combination, [...] for philosophical discussion things are clearer if we keep them separate" (ibid., 17). What might "keeping separate" mean here?

Clearly, it is not a matter of physically isolating one from the other, like severing the corpus callosum. Is the distinction then just mere word-play? On a superficial level, "feeling" and "doing" are certainly different words. However, the suggestion by Chalmers and others is that those words "cover different phenomena, both of which are quite real" (ibid., 11). Since we are dealing with something more substantive than a plurality of words (Block 2000, 133) yet less palpable than an outright physical separation, I have argued that we are confronted with what Scotus called a "formal distinction."

A formal distinction lies between a "distinction of reason" and a "real distinction." A distinction of reason "is completely dependent upon the mind" (Jordan 1984, 44), whereas "things are really distinct if they are separable, that is, if they can exist one without the other" (ibid., 45). We can use this distinction of distinctions to disambiguate important puzzles about phenomenal consciousness. Block (1995) treats the distinction between access and phenomenon as a "real" distinction in Scotus' sense. However, I submit that, if we want to separate qualities from the responses they elicit, such a real distinction is not open to us. To look for p-consciousness without a-consciousness is like looking for the smile without the Cheshire cat. But, even if the world does not permit us to encounter a quality in isolation, a formal distinction allows us to consider any quality "in its own suchness, while we disregard the connections" to anything else (Peirce 1931–58, 1.424).

Chalmers remarks that "[t]he clearest cases of direct phenomenal concepts arise when a subject attends to the quality of an experience and forms a concept wholly based on the attention to the quality" (2010, 267). Once we enlist the resources of prescissive abstraction, we can better articulate this idea of a concept *wholly* based on attention to a quality.

Some philosophers have argued that, since we employ a special class of concepts when discussing conscious states, a better understanding of those concepts will show that "the disturbing effect of the explanatory gap arises from an illusion [...]" (Loar 1999, 103). I agree that our understanding of qualia is wedded to our conceptual apparatus, but I do not see why this should make our idea of qualia illusory. Appeals to conceivability have come under attack (Yablo 1993), especially in philosophy of mind (e.g., Hill and McLaughlin 1999; Sommers 2002), so some may be uncomfortable with the idea that qualia are not (and could never be) encountered were it not for our abstract intervention. We might therefore want to distinguish "supposing" and "conceiving." In any event, I readily accept that qualia cannot be separated from functional involvement (Cohen and Dennett 2011; Churchland 2011). I nevertheless agree with Peirce that weaker grades of separation can be made. To be sure, if we reify what we prescind, our analysis will be mishandled in a way that generates insoluble problems. Still, by employing prescission, the idea of qualia can be rendered sensible.

Sydney Shoemaker provides us with a nice example of what makes prescission possible:

> If I perceive a French tricolor, I perceive a rectangle made of three horizontal stripes, of red, white, and blue. This involves experiences of those individual stripes. *There seems a good sense in which I could have had the experience I had of any of those stripes without having the experiences of the others.* (Shoemaker 2003, 65; emphasis added)

The French flag is a conjunction of three colours (in fact, the striped flag has a rectangular border, so there may be even more contrasts). We can make a real distinction by cutting out a given conjunct with scissors, but only a formal distinction allows us to truly arrive at a lone tone. Deleting all relations means deleting all comparisons. So, once there, there is no telling which flag a colour might be a part of.

Such a qualitative state is not (and could never be) encountered in actual experience, so it is, as Peirce (1998, 294) says, "a mere logical possibility." But, prescissive analysis is enough to show that philosophers of mind who believe in qualia are not discussing something entirely groundless. In the paper that propelled the contemporary debates over the "explanatory gap," Joseph Levine mentions how John Locke thought that even if "simple ideas go with their respective corpuscular configurations because God chose to so attach them [...] imagination will pry them apart" (1983, 359). If one drops the allusion to God, this begins to look like my account.

Because philosophy of signs is broader than philosophy of language, it adds something genuinely novel to the debates. Iconicity in particular lets us have a direct appreciation—or "revelation"—of qualities. By contrast, acquaintance always keeps its object at bay. Now, consider that "appearance" just means "whatever appears." Combining these notions leads current philosophers of mind into asking questions like: are we acquainted with our appearances? Philip Goff (2015, 124) feels that, to answer yes, he must posit what he calls "real acquaintance." Strawson does the same with his "real direct realism" (2015). Yet, given that acquaintance involves Secondness, it is only a matter of time before someone takes these proposals and asks: since real acquaintance is just how things appear, are we *really* acquainted with that? I prefer to sidestep this perpetual nudging away by working from a paradigm that explicitly includes Firstness in its basic commitments. I also jump to a less populated but sturdier branch in the tree of philosophy. What results is, I think, more plausible. At minimum, Peirce's semiotic analysis does justice to a fundamental truth: the idea of a relatum without a relation makes sense, but the idea of a relation without relata does not. So long as humans are capable of realizing this—and of performing the prescissive deletion which this asymmetry permits—worries about the intrinsic character of experience will persist, perhaps under different guises.

Block (2002) acknowledges that an endorsement of qualia does not fit well with scientific naturalism. Yet, catering to naturalism, he feels compelled to postulate a separate "module" in the brain that researchers would subsequently have to vindicate. The account I have developed does not make promises it cannot keep. In the story I am telling, prescission suffices to establish that a tone is subsumed in a token. However one wants to describe this analysis, Peirce was probably right when he characterized semiotics as "the quasi-necessary, or formal, doctrine of signs" (1931–58, 2.227).

Firstness is not mind, but it is not matter either. It is simply a quality, robbed of any relation(s) with anything else. This neutrality or indifference may seem contentious, but I think it can be verified using any nearby quality, like the colour of a blank sheet of paper. As Levine notes, when studying such an experience, "[w]ho can tell whether its ultimate ontological status is material or immaterial merely by means of having it?" (2001, 128). So, whereas some philosophers liken qualia to "ectoplasm" (e.g., Ladyman et al. 2007, 39–40; Majeed 2013, 254–255), I do not think the idea of a pure quality is beyond the pale.

Dennett (1991), by contrast, believes that the idea of qualia is nothing but an unfounded "meme" that has gone viral. Of course, some memes are more useful than others. So, for a time, Dennett (1987) sought to recuperate the instrumentalist benefits of discourse about "minded" creatures. However, because the phenomenal qualia at the center of recent consciousness debates could never increase our predictive success of a creature's behaviour (Ross 2005), Dennett (2006) eventually became a vocal opponent of qualia. If, like Dennett, one expects every aspect of the mind to meet Darwinian standards of increased fitness, then folding semiosis onto itself so as to inspect its incipient substructure may well be a useless ability. I do not want to go on the record as saying this, because I suspect that the intrinsic qualities which generate the hard problem of consciousness also enable diagrammatic reasoning, which is surely a useful tool (see Champagne 2016). Even so, because only sapient creatures notice (and question) their sentience, the main boon of prescissive abstraction may be that it satisfies a sense of curiosity and wonder.

People like Francisco Varela (1996) advocate a combination of third-person science and first-person phenomenology, scientists explaining the functional side and phenomenologists explaining the qualitative side. I advocate a combination of third-person science and third-person philosophy, scientists explaining everything and philosophers explaining the persistent sense that this is not enough. I think this philosophical analysis renders a worthwhile service. Nevertheless, since understanding the root cause of our perplexity regarding qualia offers no obvious technological applications, I am comfortable with the possibility that, in the end, my account of consciousness "leaves everything as it is" (Wittgenstein 2001, 42).

One of the things that my semiotic account leaves intact is the self. In the current literature on consciousness, it is sometimes thought that any theory which takes the conscious mind to be comprised of qualities must face a "combination problem" (James 2007, 158–162; Seager 1995, 280–281). The worry, in essence, is that since a self is unified, and since no intrinsic quality entails another, it is a mystery that/how a qualitative mind-stuff could ever yield a self. Some, like Strawson (1999, 100) have simply replaced unity with many little selves. This, to my mind, just makes him face many little combination problems. Unfortunately, proposals that aim for genuine unity have gotten even more bizarre (e.g., Coleman 2012; Cunningham 2013; Seager 2010). Fortunately, my account does not need these solutions, because it does not face a combination problem to begin with. Indeed, I have been insisting throughout this book that we never actually split the stream of sign-action into parts. Rather, we prescind. This requires us, at each step, to keep in mind that the starting point that we are analyzing is in fact bound. I never really took my consciousness apart, so I do not have to piece it back together.

Philosophers of signs can certainly tell an informative story about how humans construct, sustain, and reshape their self-concept and personal identity (for a promising outline, see Colapietro 1989). But, an account that explicitly uses prescission is not burdened with combining basic experiential qualities into something that looks psychologically plausible, because plausible human experience—the sort I am intimately familiar with and ardently care to enrich—has been there all along.

The stream of consciousness streams because of the action of signs. Although this action of signs is always triadic, humans can conceive—and so request an account of—the intrinsic, non-relational, nature of any experience, because we are the sorts of beings for whom that idea makes sense. It makes sense because complexity implies simplicity. Triadic signs thus subsume brute collisions and simple qualities that are not articulate (and cannot be articulated) linguistically. However, given that qualitative simplicity does not entail complexity (or any other kind of relation, save latent similarity), this asymmetry can act like a fishhook, letting us reach ineffable Firstness but preventing us from going back to Thirdness, where cognition, discourse, and science are possible. If this book has helped to show how/why philosophy of mind can sometimes get caught by that hook, then I count that as progress.

References

Block N (1995) On a confusion about a function of consciousness. Behav Brain Sci 18(2):227–287
Block N (2000) How not to find the neural correlate of consciousness. Intellectica 2(31):125–136
Block N (2002) The harder problem of consciousness. J Philos 99(8):391–425
Chalmers DJ (1996) The conscious mind: in search of a fundamental theory. Oxford University Press, Oxford
Chalmers DJ (2010) The character of consciousness. Oxford University Press, Oxford
Champagne M (2016) Brandom, Peirce, and the overlooked friction of contrapiction. Synthese 193(8):2561–2576
Churchland PM (2011) Consciousness and the introspection of "qualitative simples". Eidos 15:12–47
Cohen MA, Dennett DC (2011) Consciousness cannot be separated from function. Trends Cogn Sci 15(8):358–364
Colapietro VM (1989) Peirce's approach to the self: a semiotic perspective on human subjectivity. State University of New York Press, Albany
Coleman S (2012) Mental chemistry: combination for panpsychists. Dialectica 66(1):137–166
Crane T (2000) The origins of qualia. In: Crane T, Patterson S (eds) History of the mind-body problem. Routledge, London, pp 169–194
Cunningham PF (2013) Explaining consciousness: a (very) different approach to the "hard problem". The Journal of Mind and Behavior 34(1):41–62
Dennett DC (1987) The intentional stance. MIT Press, Cambridge
Dennett DC (1991) Consciousness explained. Little, Brown and Company, New York
Dennett DC (1998) Brain children: essays on designing minds. MIT Press, Cambridge
Dennett DC (2006) Sweet dreams: philosophical obstacles to a science of consciousness. MIT Press, Cambridge
Goff P (2015) Real acquaintance and physicalism. In: Coates P, Coleman S (eds) Phenomenal qualities: sense, perception, and consciousness. Oxford University Press, Oxford, pp 121–146
Hill CS, McLaughlin BP (1999) There are fewer things in reality than are dreamt of in Chalmer's philosophy. Philos Phenomenol Res 59(2):445–454
James W (2007) The principles of psychology, vol 1. Cosimo, New York
Jordan MJ (1984) Duns Scotus on the formal distinction. Dissertation, Rutgers University
Ladyman J, Ross D, Spurrett D, Collier J (2007) Every thing must go: metaphysics naturalized. Oxford University Press, Oxford
Levine J (1983) Materialism and qualia: the explanatory gap. Pac Philos Q 64(4):354–361
Levine J (2001) Purple haze: the puzzle of consciousness. Oxford University Press, Oxford
Livingston PM (2004) Philosophical history and the problem of consciousness. Cambridge University Press, Cambridge
Loar B (1999) Should the explanatory gap perplex us? In: Rockmore T (ed) Proceedings of the twentieth world congress of philosophy, vol 2. Philosophy Documentation Center, Bowling Green, pp 99–104
Majeed R (2013) Pleading ignorance in response to experiential primitivism. Philos Stud 163(1):251–269
Peirce CS (1931–58) The collected papers of Charles Sanders Peirce. Harvard University Press, Cambridge
Peirce CS (1998) The essential Peirce: selected philosophical writings, vol 2. Indiana University Press, Bloomington
Ross D (2005) Chalmers's naturalistic dualism: a case study in the irrelevance of the mind-body problem to the scientific study of consciousness. In: Erneling CE, Johnson DM (eds) The mind as scientific object: between brain and culture. Oxford University Press, Oxford, pp 165–175
Seager WE (1995) Consciousness, information, and panpsychism. J Conscious Stud 2(3):272–288
Seager WE (2010) Panpsychism, aggregation and combinatorial infusion. Mind and Matter 8(2):167–184

Shoemaker S (2003) Consciousness and co-consciousness. In: Cleeremans A (ed) The unity of consciousness: binding, integration and dissociation. Oxford University Press, Oxford, pp 59–71
Sommers T (2002) Of zombies, color scientists, and floating iron bars. Psyche 8(22)
Strawson G (1999) The self and the SESMET. J Conscious Stud 6(4):99–135
Strawson G (2015) Real direct realism: reflections on perception. In: Coates P, Coleman S (eds) Phenomenal qualities: sense, perception, and consciousness. Oxford University Press, Oxford, pp 214–253
Varela FJ (1996) Neurophenomenology: a methodological remedy for the hard problem. J Conscious Stud 3(4):330–349
Wittgenstein L (2001) Philosophical investigations (trans: Anscombe GEM). Blackwell, Oxford
Yablo S (1993) Is conceivability a guide to possibility? Philos Phenomenol Res 53(1):1–42

Index

M. Champagne, *Consciousness and the Philosophy of Signs*, Studies in the History of Philosophy of Mind 19, https://doi.org/10.1007/978-3-319-73338-8

The manufacturer's authorised representative in the EU is Springer
Nature Customer Service Centre GmbH, Europaplatz 3, 69115 Heidelberg,
Germany. If you have any concerns regarding our products, please
contact ProductSafety@springernature.com

Printed and bound by CPI Group (UK) Ltd, Croydon, CR0 4YY
30/09/2026
02225606-0012